Solar Photovoltaic Cells
Photons to Electricity

Solar Photovoltaic Cells
Photons to Electricity

Alexander P. Kirk
Arizona State University
Tempe, Arizona, USA

AMSTERDAM • BOSTON • HEIDELBERG • LONDON
NEW YORK • OXFORD • PARIS • SAN DIEGO
SAN FRANCISCO • SINGAPORE • SYDNEY • TOKYO
Academic Press is an imprint of Elsevier

Academic Press is an imprint of Elsevier
32 Jamestown Road, London NW1 7BY, UK
525 B Street, Suite 1800, San Diego, CA 92101-4495, USA
225 Wyman Street, Waltham, MA 02451, USA
The Boulevard, Langford Lane, Kidlington, Oxford OX5 1GB, UK

Library of Congress Cataloging-in-Publication Data
A catalog record for this book is available from the Library of Congress

British Library Cataloguing-in-Publication Data
A catalogue record for this book is available from the British Library

For information on all Academic Press publications
visit our website at http://store.elsevier.com/

ISBN: 978-0-12-802329-7

Working together
to grow libraries in
developing countries

ELSEVIER   Book Aid International

www.elsevier.com • www.bookaid.org

DEDICATION

To Joan

CONTENTS

PREFACE

Only natural sunlight with its polychromatic splendor is capable of properly illuminating and revealing the beautiful and subtle colors, shapes, and textures of our environment – fields of flowers, forests, mountain peaks, waterfalls, seashores, or desert sands. And, sunlight really dazzles us with rainbows after a rain shower. All of this magnificent imagery alone is enough to compel and inspire us to ponder sunlight and wonder about ways to harness its energy. With this thought in mind, this book was written to provide an understanding of how sunlight can be used by humans to give clean electricity enabled by a fascinating solid-state energy conversion device known as the solar photovoltaic cell.

This compact book has been written for anyone interested in learning about solar photovoltaic cells. Hopefully it will succinctly reveal the usefulness of solar electricity and the beauty of solar photovoltaic cells while outlining the benefits and challenges associated with solar photovoltaic technology and widely scaled solar electricity generation.

The author is grateful for his mentor, Prof. Yong-Hang Zhang at Arizona State University, for leading numerous stimulating, enlightening, and valuable in–depth discussions on the important attributes and governing factors that are required in order to attain the most efficient and highest performance solar photovoltaic cells. The author extends his gratitude to Dr. Christian A. Gueymard at Solar Consulting Services and Prof. David K. Ferry at Arizona State University for reviewing portions of this book and offering clarifications, corrections, and suggestions for improvement. The author acknowledges Dr. Pablo Benítez at Universidad Politécnica de Madrid for helpful comments about CPV optics. The author wishes to thank his father, Prof. Wiley P. Kirk at University of Texas at Arlington, for not only many helpful comments on this book but for catalyzing his interest in solar photovoltaic cells and the science behind solid-state energy conversion devices to begin with. Additionally, the author thanks the Arizona State University Noble Science Library staff for help accessing reference books and journal articles. The author acknowledges Lisa Reading, Natasha Welford, and Anusha Sambamoorthy at Elsevier (Academic Press) for enabling this book to come to fruition and for being most helpful and a joy to work with. Finally, the author acknowledges Science Foundation Arizona's Bisgrove Scholars program and Dr. Mary O'Reilly at the foundation for her enthusiasm and encouragement. This book could not have been written without the generous support of this scholarship while the author was at Arizona State University. The opinions, findings, and conclusions or recommendations expressed in this material are those of the author and do not necessarily reflect the views of the Science Foundation Arizona.

A. P. Kirk
Tempe, AZ, USA

INTRODUCTION

The overarching goal of this book on solar photovoltaic cells is:

1. to motivate their existence and rationale for deployment
2. to examine the sunlight that irradiates them
3. to explain their operation
4. to benchmark their performance to combined-cycle thermal power plants
5. to investigate semiconductor raw material demands as well as issues related to scalability

The natural connection we humans already have with sunlight in our daily lives is presented in Chapter 1 where we consider how much energy we demand not just to remain alive but also to live life in our increasingly technology-dense societies. From this, the electricity demand in the USA and also globally is examined, and the amount of electricity generated from photovoltaic modules is compared to other electricity generation technologies such as coal-fired power plants and wind turbines.

Chapter 2 begins with the nuclear fusion reactions in our Sun and builds into a discussion of the solar radiation we ultimately receive here on Earth. Topics of interest include total solar irradiance and the extraterrestrial solar spectrum, air mass and atmospheric attenuation of sunlight, direct and global terrestrial solar spectra including their modeling, and solar photon flux.

A step-by-step presentation of the physics and operating characteristics of solar photovoltaic cells is presented in Chapter 3 including topics such as the relationship between bandgap energy and power conversion efficiency, detailed balance, photogenerated current density, open circuit voltage, free energy, hot carrier relaxation, and multiple junction cell architecture and its benefits.

The concept of energy cascading is invoked in Chapter 4 as a framework to compare high-performance six junction solar photovoltaic cells operating under concentrated sunlight with combined-cycle thermal power plants comprised of natural gas and steam turbines. Throughout this case study, calculations of theoretical limiting cell power conversion efficiency as well as estimates of achievable AC module efficiency are presented in order to understand not just the limits but also the capabilities of advanced multijunction cells operating under concentrated sunlight.

The purpose of Chapter 5 is to offer a brief examination of the potential scalability of solar photovoltaic cells and modules based on the following semiconductors: Si, CdTe, and CIGS (cells operating under conventional one Sun illumination) as well as Ge and GaAs (cells operating under concentrated sunlight). To a first order approximation, the amount of precious semiconductor material that will be required to achieve a relatively modest peak power output of one terawatt is investigated. Finally, soft costs, solar energy storage, and the evolution of the electric grid are discussed.

An image gallery is included near the end of the book in Chapter 6 to highlight some of the real technologies that make solar electricity possible.

CHAPTER *1*

Energy Demand and Solar Electricity

1.1 INTRODUCTION

This chapter begins with the connection between humans on Earth and the Sun that we depend on, and must adapt to, in order to survive. Then, we quantify how much energy is needed on average to sustain human life. This human energy demand is used as a benchmark when we next investigate electricity generation in the USA followed by global electricity generation. Solar electricity generation is compared with other electricity generation sources such as wind, natural gas, and coal.

1.2 HUMAN-SUNLIGHT CONNECTION

The remarkable human eye contains photoreceptor cone cells that respond to the visible portion of the solar spectrum from ~380 nm violet light to ~740 nm red light with photopic (i.e., bright light) vision peak sensitivity corresponding to ~555 nm green light [1], as shown in Figure 1.1. The near-ultraviolet light that is not absorbed by atmospheric ozone is used by our skin to synthesize vitamin D.

Meanwhile, oxygenic photosynthetic land plants (via chlorophyll and carotenoid molecules) effectively use visible light from the Sun to produce carbohydrate and perhaps the most benign of all waste products – pure oxygen that we breathe. The overall oxygenic photosynthesis process may be expressed by the chemical reaction given by $H_2O + CO_2 \rightarrow O_2 + CH_2O$, where H_2O is water, CO_2 is carbon dioxide, O_2 is oxygen, and CH_2O is used here to represent a generic carbohydrate subunit.

Sunlight-dependent plants provide us plant-dependent humans with the food (carbohydrate) and oxygen that we need to survive and thrive. Plants provide us with clothing (e.g., cotton and linen), shelter for us (e.g., framing material from pine or thatched roofs from straw) and habitat for other animals, shade, furniture, cabinetry, decking and flooring,

Solar Photovoltaic Cells: Photons to Electricity. http://dx.doi.org/10.1016/B978-0-12-802329-7.00001-8

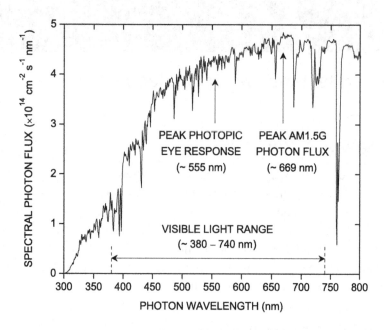

Fig. 1.1. Peak photopic eye response with respect to AM1.5G photon flux and visible light. Note: the AM1.5G terrestrial solar spectrum (developed by Gueymard) and solar photon flux will be described in Chapter 2.

shipping crates, medicines and homeopathic remedies, paper and card-board products such as books and boxes, dyes and pigments, perfumes and fragrances, cooking oils, delicious beverages such as tea and coffee, a sink for carbon dioxide, filtration of airborne pollutants, mitigation of soil erosion, coastal storm surge buffering, sound damping, wind breaks, sports fields, and simply enjoyment in our gardens, parks, and nature preserves.

Moreover, the Sun provides warmth, keeps in operation the hydro-logic cycle (evaporation of water and precipitation), and in large mea-sure dictates the Earth's seasonal climate and weather that we must respond and adapt to. Humans even tailor their accessories to enable functionality in sunlight, for example, by using sunglasses and hats or by designing clothing for sunny days such as colorful women's sun-dresses. With our natural connection to the Sun and solar radiant energy, it is inherently logical for humans to use sunlight for the purpose of generating clean electricity through the application of solar photo-voltaic cells.

1.3 HUMAN ENERGY REQUIREMENT

Each day, considering an average value, humans require ~2.33 kWh (kilowatt–hours) of chemical energy to live a healthy life. Typically, instead of the units of kWh, this is expressed (in the USA) on food packaging labels as kcal (kilocalories) where 2.33 kWh is about equal to 2000 kcal [2]. Therefore, each month, on an average, humans require ~70 kWh of energy. And, in one year, this equates to ~850 kWh. Stated another way, powering a 100 W light bulb 24 h a day requires about the same energy as the human body each day. With ~7.1 × 10⁹ humans on the Earth in 2013 [3]; this is nearly equivalent to continually operating a quantity of 7.1 × 10⁹ of the 100 W light bulbs, which results in a yearly energy expenditure of ~6 × 10¹² kWh or 6 PWh (petawatt–hours). Next, we will compare the demand for (chemical) energy that is required simply to power our bodies and sustain our lives to the demand for electricity that we then use to enrich our lives in technology-dense societies such as the USA.

1.4 ELECTRICITY GENERATION IN THE USA

In 2013, total net generation of electricity ("all sectors") in the USA was ~4.06 × 10¹² kWh [4]. At the end of 2013, the population of the USA was ~3.17 × 10⁸ [5]. If we normalize electricity consumption in the year 2013 by the population, we find that this leads to an equivalent of ~1.3 × 10⁴ kWh per year of electricity per person to power the USA and run its economy. This means that ~35 kWh per day per person of electricity was utilized on an average, or ~15× more energy per day than it takes just to sustain a human body in a state of good health (~2.33 kWh per day) as we found in the preceding section. Therefore, on the one hand, while we need a sustainable agricultural base just to maintain human life by providing enough food (chemical energy), on the other hand, we see that living in a technology-dense society requires much more energy if we want to power industrial machinery and processes, run air conditioners and heat pumps, operate computers and charge mobile phone batteries, turn on the lights, and so forth.

Most of our electricity comes from thermal power plants that require either the combustion of fossil fuels or fission of radioactive materials.

Specifically, the heat from combustion of coal or fission of uranium is used to convert water into steam; the steam is expanded in a steam turbine that in turn is connected to a generator that converts rotational mechanical energy into alternating current (AC) electricity. Natural gas may be combusted in a gas turbine that is coupled to a generator to generate AC electricity. In Chapter 4, we will discuss combined-cycle power plants that incorporate a gas turbine and a steam turbine that utilizes the waste heat from the gas turbine exhaust.

In 2013, electricity generated by the combination of coal, natural gas, and nuclear fission power plants yielded ~86% of the total electricity net generation in the USA [4]. Therefore, this is where the motivation for solar photovoltaic cells comes from – the ability to generate electricity by using sunlight now, instead of the nearly complete reliance on fossil fuels such as coal and natural gas or fissile radioactive materials such as uranium (^{235}U). As shown in Figure 1.2, solar photovoltaic electricity generation has recently been expanding rapidly in the USA. Nonetheless, in 2013, solar photovoltaic electricity net generation ("all sectors") was ~8.3×10^9 kWh, as shown in Figure 1.2, or only ~0.2% of the total

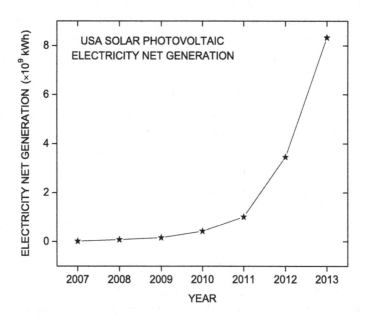

Fig. 1.2. Solar photovoltaic electricity net generation trend in the USA from 2007 to 2013. (Data from Ref. [4].)

electricity net generation (all sectors) in the USA. This is actually much less even than the ~7% per year waste of electrical energy (i.e., a staggering 2.79 × 10¹¹ kWh in 2013) through losses associated with transmission and distribution [4].

There are a number of other ways to generate electricity including the use of wind turbines, hydroelectric plants, fuel cells, thermoelectric generators, and concentrated solar power – a solar thermal technology. As an example, in 2013, wind and hydroelectric, in particular, provided ~4.1% and ~6.6%, respectively, of the total electricity net generation in the USA [4].

For more clarity, solar photovoltaic electricity net generation in the USA is shown versus geothermal, wood biomass, wind, hydroelectric, nuclear (fission), natural gas, and coal electricity in Figure 1.3. Note here that the data in this section are from the U.S. Energy Information Administration's *Monthly Energy Review, 7.1 Electricity Overview* and *7.2a Electricity Net Generation: Total (All Sectors)* and *Electric Power Monthly, Table 1.1.A. Net Generation from Renewable Sources: Total (All Sectors)*. Periodic updates occur and so there may be revisions to the data [4].

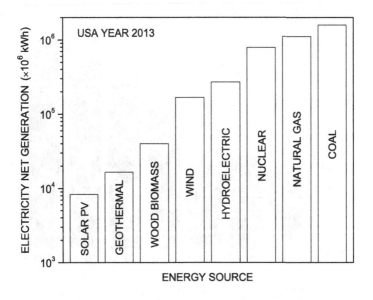

Fig. 1.3. Comparison of electricity net generation by energy source in the USA in 2013. (Data from Ref. [4].)

Before moving on, we will briefly discuss the combustion of fossil fuels for the purpose of generating electricity. Just in the USA in 2013 alone, there was ~7.8 × 10^{11} kg of coal (anthracite, bituminous, subbituminous, lignite, waste coal, as well as coal synfuel) consumed for generating electricity for a country that represents roughly 4% of the total world population. The combustion of coal leads to exhaust emission that contains CO_2, mercury (Hg), and sulfur dioxide (SO_2). Fly and bottom ash is also accumulated. The combustion of coal in the electric power sector in the USA in 2013 resulted in the estimated emission of ~1.6 × 10^{12} kg of CO_2. By comparison, in 2013 in the USA, there was ~7.8 × 10^{12} ft^3 of natural gas consumed for electricity generation resulting in ~4.4 × 10^{11} kg of CO_2 emissions. The data here are from the U.S. Energy Information Administration's *Monthly Energy Review*, specifically *Table 7.3b Consumption of Combustible Fuels for Electricity Generation: Electric Power Sector* and *Table 12.6 Carbon Dioxide Emissions from Energy Consumption: Electric Power Sector* [4].

1.5 GLOBAL ELECTRICITY GENERATION

In 2013, global electricity generation (gross output) was ~2.3 × 10^{13} kWh, as shown in Figure 1.4.

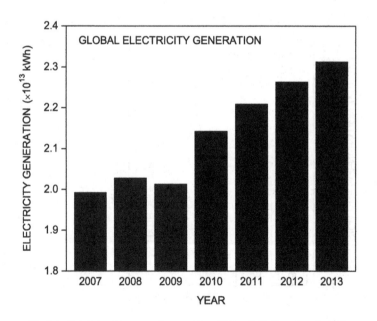

Fig. 1.4. Global electricity generation trend from 2007 to 2013. (Data from Ref. [6].)

Meanwhile, in 2013, global solar electricity generation was only $\sim 1.25 \times 10^{11}$ kWh, as shown in Figure 1.5, which represents $\sim 0.5\%$ of the global total electricity generation. As of 2013, the top ten solar electricity generating countries were Germany, Italy, Spain, China, Japan, USA, France, Australia, Czech Republic, and UK, as shown in Figure 1.6. Data here are from BP's *Statistical Review of World Energy 2014* [6].

Whereas European countries that are relatively deficient in solar irradiance, such as Germany, Czech Republic, and UK, are in the top ten solar electric power generators, countries with abundant solar irradiance in the Middle East (e.g., Saudi Arabia), Africa (e.g., Egypt), North America (e.g., Mexico), and South America (e.g., Chile) so far only have limited solar electricity generation. Even the USA and China – countries with abundant solar irradiance – are only generating a trivial fraction of electricity from sunlight. What this means in essence is that trillions upon trillions of solar photons that could have been used to generate clean electricity are instead just being absorbed in places such as rooftops, parking lots, railway and road easements, and agriculturally/ecologically marginal lands where photovoltaic modules could have been deployed.

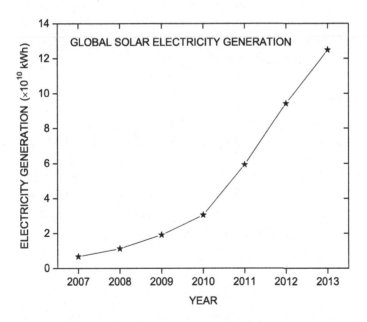

Fig. 1.5. Solar (photovoltaic + thermal) electricity generation trend globally from 2007 to 2013. (Data from Ref. [6].)

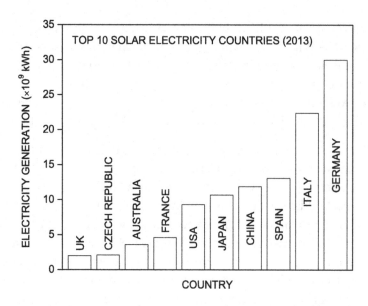

Fig. 1.6. Top 10 countries for solar electricity generation in 2013. (Data from Ref. [6].)

At least in the next few decades, global population is forecast to increase along with the demand for electricity. For example, the United Nations (UN) predicts the global population will be $\sim 8.4 \times 10^9$ ("medium variant" estimation) by 2030 [3] while BP's *Energy Outlook 2035* predicts that global electricity generation will be $\sim 3.3 \times 10^{13}$ kWh by 2030 [6].

REFERENCES

[1] S.M. Sze, Physics of Semiconductor Devices, 2nd ed., John Wiley & Sons, New York, 1981.

[2] P. Würfel, Physics of Solar Cells: From Basic Principles to Advanced Concepts, 2nd ed., Wiley-VCH, Weinheim, 2009.

[3] United Nations. Available from: http://www.un.org/en/development/desa/population/.

[4] Energy Information Administration. Available from: http://www.eia.gov/totalenergy/data/monthly/.

[5] United States Census Bureau. Available from: http://www.census.gov/popclock/.

[6] BP. Available from: http://www.bp.com.

CHAPTER 2

From Nuclear Fusion to Sunlight

2.1 INTRODUCTION

This chapter begins with the mass and composition of the Sun and then evolves into a simple introduction of nuclear fusion and what is known as the proton–proton (*pp*) chain. Next, a brief overview of solar neutrinos is presented, followed by a brief discussion of quantum mechanical tunneling of the hydrogen proton wave functions through the repulsive Coulomb barrier. From here, the luminosity of the Sun is discussed. Then, the different regions of the Sun are outlined as well as the time it takes for sunlight (photons) to reach the Earth. The chapter continues by giving a discussion of total solar irradiance (TSI) and how TSI is measured. From this, the discussion focuses on the extraterrestrial (AM0) solar spectrum, and then there is a discussion of air mass and how this is calculated. Next, atmospheric aerosols and scattering are discussed including aerosol optical depth (AOD) followed briefly by a mention of clouds. Then, both direct beam and global terrestrial solar spectra are discussed. Finally, photon flux, which will be used later in this book to compute the photogenerated current density of solar photovoltaic cells, is presented. Overall, the primary goal of this chapter is to outline the following: processes in the Sun that make it possible for sunlight to be radiated into space, attenuation of sunlight by Earth's atmosphere, direct and global terrestrial solar irradiance, and the photon flux that is absorbed by solar photovoltaic cells and converted into photocurrent, which will be discussed in Chapter 3.

2.2 THE MASSIVE SUN

Our Sun, already $\sim 4.6 \times 10^9$ years old [1], is presently a main sequence star meaning that it is in a period of its life whereby there is still hydrogen, or nuclear fuel, in the central core that has not been fused into helium. While Earth, a rocky planet with an iron core [2], has a mass of $\sim 5.97 \times 10^{24}$ kg, the mass of the Sun today is $\sim$333,000$\times$ greater at $\sim 1.99 \times 10^{30}$ kg. This mass value is impressive considering that the Sun, a plasma, is almost entirely just comprised of the two lightest elements, $\sim$71% hydrogen and $\sim$27% helium with $\sim$2% remaining mass from elements such as carbon, nitrogen, and oxygen among others (all values are given as mass fraction at zero age main sequence) [1]. It is worth noting that the Sun now is thought to be in what is known as a "hydrostatic equilibrium" which means that gravity provides a counterbalance to the particle and radiative pressure [1,3].

2.3 NUCLEAR FUSION SEQUENCE

In the hot ($\sim 1.5 \times 10^7$ K) central core of the Sun, hydrogen nuclei (protons) form helium by nuclear fusion [1] (not to be confused with nuclear fission). Nuclear fusion in the Sun proceeds via steps in what is known as the *pp* chain [3–5]. The first step [6] is described by:

$$^1H + {}^1H \rightarrow {}^2H + e^+ + v_e, \tag{2.1}$$

where ^1H represents a proton, ^2H represents a deuteron, e^+ represents a positron, and v_e represents an electron neutrino known specifically here in Equation 2.1 as the *pp* neutrino. The second step is described by:

$$^2H + {}^1H \rightarrow {}^3He + \gamma, \tag{2.2}$$

where ^3He represents a nucleus of light helium and γ represents a gamma-ray photon. The third step involves the fusion of two ^3He nuclei to produce a nucleus of ^4He (α-particle) as described by:

$$^3He + {}^3He \rightarrow {}^4He + 2\,^1H. \tag{2.3}$$

This third step occurs $\sim$83.3% of the time in the Sun. In summary, these reactions are often expressed as:

$$4\,^1H \rightarrow {}^4He + 2e^+ + 2v_e. \tag{2.4}$$

About 16.7% of the time, instead of the third step already shown, the following reaction occurs instead: $^3He + {}^4He \rightarrow {}^7Be + \gamma$, which either leads $\sim$99.88% of the time to $^7Be + e^- \rightarrow {}^7Li + v_e$; $^7Li + {}^1H \rightarrow 2\,^4He$ where the neutrino here is known as the ^{7}Be neutrino or $\sim$0.12% of the time to ^{7}Be + ^{1}H $\rightarrow$ ^{8}B + γ; ^{8}B $\rightarrow$ ^{8}Be* + e^+ + ν_e; ^{8}Be* $\rightarrow$ 2^4He where the neutrino here is known as the ^{8}B neutrino. In addition, there is the rare reaction ($\sim$2 $\times$ 10^{-5}% of the time) given by ^{3}He + ^{1}H $\rightarrow$ ^{4}He + e^+ + ν_e, where the neutrino here is known as the *hep* neutrino. As a side note, positrons and electrons annihilate (e^+ + e^- $\rightarrow$ 2γ).

Before continuing, and just for added clarity, instead of step 1 (shown in Eq. 2.1) that occurs $\sim$99.76% of the time, about $\sim$0.24% of the time the reaction given by ^{1}H + e^- + ^{1}H $\rightarrow$ ^{2}H + ν_e will occur. This is known as the proton–electron–proton (*pep*) reaction and the resultant electron neutrino here is known as the *pep* neutrino. The first detection of *pep* neutrinos was made at the Borexino detector in Italy in 2012 [7].

2.4 SOLAR NEUTRINOS

Diverging slightly for the moment, the solar electron neutrinos are quite interesting in their own right because they are thought to change "flavor" (a quantum mechanical process) in transit through the Sun and the Earth, sometimes oscillating from an electron neutrino variant to a tau or muon neutrino variant. As of the year 2014, it is believed that neutrinos have a vanishingly small mass although the mass values of the three different flavors of neutrinos are not exactly known. Neutrinos travel through the Sun, through space, through the Earth, and through us in remarkable numbers [1,3].

It takes specialized deep underground detectors at places such as Homestake Mine in the USA, Borexino in Italy, Sudbury Neutrino Observatory (SNO) in Canada, Super-Kamiokande (SK) in Japan, and IceCube in the Antarctic to properly detect neutrinos. Neutrinos are uncharged fermions (leptons) and are not actually observed directly. For example, in the SK detector that contains a massive stainless steel tank filled with 50 $\times$ 10^3 tons of ultrapure water located 1000 m underground in an old mine [8], a ^{8}B neutrino may, in a rare event, collide with an electron in a water molecule in an elastic scattering process. As the recoiling electron (a charged particle) moves with velocity in water faster than the phase velocity of light in water (but not faster than light travels in vacuum), Čerenkov radiation [9] is emitted in a cone shaped pattern that may be detected by photomultiplier tubes (PMT) surrounding the huge tank of ultrapure water. Ultimately, gaining a more complete understanding of neutrinos may then allow for a concomitantly more complete understanding of our Sun and possibly enrich our overall knowledge of particle physics. Toward this end, a deeper understanding of *pp* neutrinos has been published [5].

2.5 QUANTUM MECHANICAL TUNNELING

The nuclear fusion process itself is also quite remarkable because one would initially assume that two protons would stridently repel each other due to Coulomb repulsion [1,3]. It is understood, however, that despite the tendency for strong repulsion of the positively charged protons, quantum mechanical tunneling (of the wave functions of the interacting protons) through the Coulomb barrier can occur as explained by Gamow [10]. The slow rate of the nuclear reaction in step 1 (Eq. 2.1) of the *pp* chain reaction sequence allows for the controlled and long-term (billions of years) duration of nuclear fusion in the Sun [1,3].

2.6 RADIANT POWER

There are indeed a tremendous number of nuclear fusion events in the core of our Sun [1]. Each second, a staggering amount of $\sim 3.83 \times 10^{26}$ J of energy is released. Since 1 J s^{-1} = 1 W, this is 3.83×10^{26} W, which is known as the luminosity (radiant power) of the Sun. For context, a typical incandescent light bulb (which, like the Sun, also emits polychromatic or white light) is rated at 100 W, a difference of 24 orders of magnitude or a factor of a trillion-trillion [1]. If one incandescent light bulb (noting that incandescent bulbs are being phased out in favor of light emitting diodes) could be manufactured per second, it would take $\sim 10^{17}$ years to manufacture enough 100 W bulbs to reach the same radiant power as the Sun (assuming we even had the means to power all these bulbs). In comparison, the Earth is only $\sim 4.5 \times 10^{9}$ years old. Moreover, if each incandescent light bulb used 0.01 g of tungsten (W) for the active light emitting filament, then $\sim 10^{16}$ metric tons of tungsten, a relatively rare metal, would be required to match the luminosity of the Sun. As of the year 2014, known reserves of tungsten on Earth fall far short of this at an estimated amount of $\sim 3.5 \times 10^{6}$ metric tons [11]. The overall point to be made here is that the sustained radiant power of the Sun is truly phenomenal and also difficult to comprehend. Meanwhile, scientists and engineers are working to achieve controlled and sustained fusion on Earth at the National Ignition Facility (NIF) [12] and the International Thermonuclear Experimental Reactor (ITER) [13].

2.7 FROM CORE TO PHOTOSPHERE

The temperature in the core of the Sun ($\sim 1.5 \times 10^7$ K) may be approximated from:

$$T_{\text{core}} = 2Gm_{\text{S}}m_{\text{p}} / 3k_{\text{B}}R_{\text{S}}, \qquad (2.5)$$

where G is the gravitational constant (6.6738×10^{-11} N m^2 kg^{-2}), m_{S} is the mass of the Sun (1.99×10^{30} kg), m_{p} is the mass of a proton (1.6726×10^{-27} kg), k_{B} is Boltzmann's constant (1.3806×10^{-23} J K^{-1}), and R_{S} is the radius of the Sun (6.96×10^8 m). The energy released from the nuclear fusion process in the $\sim 1.5 \times 10^7$ K core is initially radiated outward from the core in a region of the Sun known as the radiative zone. The next most outward region of the Sun is known as the convective zone. The temperature at the radiative/convective zone boundary region is thought to be $\sim 1.5 \times 10^6$ K [1], or a factor of 10 cooler than the core temperature. From the convective zone, the energy transports convectively to the next most outward region known as the photosphere that has an effective temperature of ~ 5772 K, as shown in Figure 2.1. The

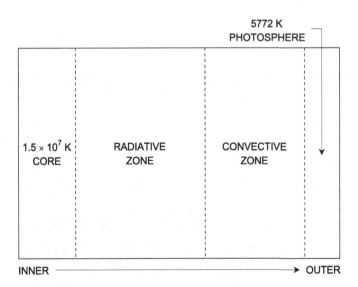

Fig. 2.1. Simplified schematic representation of different regions in the Sun.

sunlight we see here on the Earth is represented by the radiation emitted from the photosphere. Outward of the photosphere is the chromosphere and then a transition region followed even more outward by the corona. The chromosphere and corona can only be observed by eye during total solar eclipses, and at this time, the thin chromosphere, in particular, appears as a red color due to the characteristic photon emission from hydrogen-alpha (Hα) electron transitions at 656.3 nm wavelength known also as an emission line in the Balmer series [1].

2.8 LONG-DISTANCE TRAVEL

The Sun is ~1.496 × 10^{11} m from Earth [1]. This value is the average distance between the Earth and the Sun, and is known as the astronomical unit (AU). From the velocity of light in the relative vacuum of space (2.9979 × 10^8 m s^{-1}), photons radiated from the Sun in the direction of the Earth only take ~500 s (8.3 min) to reach us. In comparison, mediocre runners may take 8.3 min to run just 1.6 × 10^3 m (about 1 mile) while elite middle distance runners are able to run a little over twice this distance in the same time. Ultimately, solar energy is radiated away from the Sun in a continuous electromagnetic (EM) spectrum, as depicted in Figure 2.2.

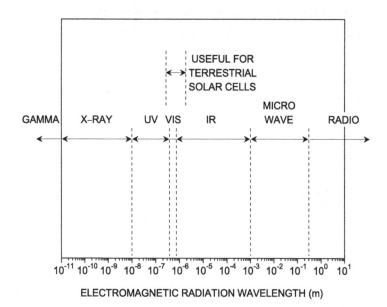

Fig. 2.2. EM radiation spectrum.

2.9 TSI

Just above the Earth's atmosphere, TSI that varies day by day has been measured by radiometers onboard various satellites since 1978. In 2003, the 290 kg Solar Radiation and Climate Experiment (SORCE) satellite sponsored by the National Aeronautics and Space Administration (NASA) was launched in a ~645 km low Earth orbit. The SORCE satellite is equipped with the total irradiance monitor (TIM) to continue the process of measuring TSI [14]. The TIM contains four electrical substitution radiometers (ESR) with design attributes such as high precision aluminum apertures and etched Ni-P (nickel-phosphorus) black internal cavity coating [15].

Due to magnetic variability, the Sun is on an approximate 11 years primary solar cycle (with minima in TSI ~11 years apart). Solar faculae (bright spots) and solar flares typically act to increase TSI, whereas sunspots typically act to reduce TSI [16]. Solar radiation, therefore, is not truly constant since noticeable fluctuations are observed day to day as shown in Figure 2.3. After 10 years of daily TSI measurements onboard the aforementioned SORCE satellite, the average TSI from

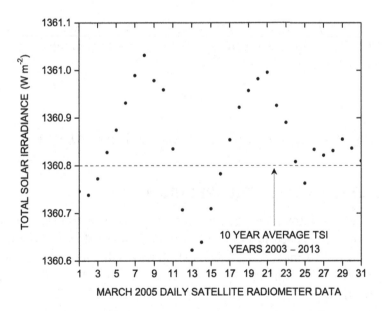

Fig. 2.3. Daily TSI measured in March 2005 by radiometers onboard the SORCE satellite. Ten-year average TSI is also shown. Data (March 2003 to March 2013) from the SORCE mission website, http://lasp.colorado.edu/home/sorce/.

March 2003 to March 2013 was $TSI_{avg} \sim 1360.8$ W m^{-2}. A caveat here is that long-term (and even more accurate) measurements of TSI are likely to result in a revised value of TSI_{avg}. For example, a new satellite launched in November 2013, known as the total solar irradiance calibration transfer experiment (TCTE), contains an updated TIM for measuring TSI [14].

One interesting point of commentary here is that the SORCE satellite is powered by solar photovoltaic cells (and batteries during orbital eclipse). To obtain accurate measurements of solar irradiance, radiant energy from the Sun is converted by solar photovoltaic cells into electricity to power the satellite that provides the platform for radiometer measurements of sunlight in the first place. Therefore, in a beautifully efficient and synergistic way, sunlight is utilized to help measure sunlight.

To close the loop on this discussion, it is actually the average TSI (~ 1360.8 W m^{-2}) that was used to calculate the aforementioned luminosity ($L \sim 3.83 \times 10^{26}$ W) of the Sun from the relationship given by:

$$L = 4\pi TSI_{avg} (AU)^2. \tag{2.6}$$

The effective "surface" temperature (~ 5772 K photosphere) of the "black body" Sun that was mentioned earlier was computed from the well-known Stefan–Boltzmann law [1]:

$$T_{eff} = (L / 4\pi\sigma R_S^2)^{1/4}, \tag{2.7}$$

where σ is the Stefan–Boltzmann constant (5.67×10^{-8} J m^{-2} K^{-4} s^{-1}).

2.10 EXTRATERRESTRIAL SPECTRUM

In the solar photovoltaics community, the solar spectrum in the relative vacuum of space just above the Earth's atmosphere is commonly known as the air mass zero (AM0) extraterrestrial spectrum that has an integrated total irradiance equivalent to the average TSI [17]. This can be expressed as:

$$I_{AM0} = TSI_{avg} = \int_0^\infty I_\lambda \, d\lambda, \tag{2.8}$$

where I_{AM0} is the total integrated irradiance (at normal incidence) of the extraterrestrial AM0 solar spectrum, TSI_{avg} is the average TSI (at normal incidence), I_λ is the spectral irradiance (at normal incidence), and λ is photon wavelength. The extraterrestrial AM0 spectrum is not only used for guiding the design of high-efficiency space solar photovoltaic cells deployed on satellites and space stations, but it is also used as the starting point to accurately model terrestrial solar spectra using established radiative transfer equations. The extraterrestrial AM0 spectrum, derived from Gueymard [16], is shown in Figure 2.4 (plotted out to 2500 nm here instead of the usual 4000 nm cutoff). The spectral irradiance characteristic of the AM0 spectrum is primarily related to the effective "surface" temperature of the Sun.

2.11 RELATIVE AIR MASS

In the context of solar energy and terrestrial solar spectra, the term "relative air mass" represents the ratio of the path length of sunlight through Earth's atmosphere to the vertical path length. It is difficult to precisely calculate the relative air mass on account of some uncertainties such as variation of atmospheric refractive index with air pressure.

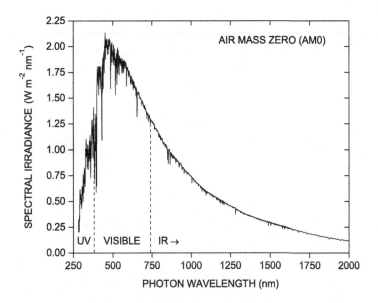

Fig. 2.4. Extraterrestrial AM0 solar spectrum. Note that the spectrum is plotted out to 2000 nm instead of the full 4000 nm cutoff. AM0 spectrum data is derived from Gueymard and scaled to an average TSI value of 1360.8 W m⁻².

Nonetheless, a suitable expression from Gueymard [18] for the approximate relative air mass (AM) is given by:

$$AM \approx [\cos\theta + 0.48353\theta^{0.095846} / (96.741 - \theta)^{1.754}]^{-1}, \quad (2.9)$$

where θ is zenith angle. It is worth mentioning that the relative air mass we are discussing here refers to the optical mass for Rayleigh scattering and mixed-gas absorption. (While relative air mass is preferred, note that the "absolute air mass" may be calculated by multiplying the right hand side of Equation 2.9 by P/P_0, where P is site atmospheric pressure in mbar and P_0 is standard atmospheric pressure, 1013.25 mbar.) When the Sun is directly overhead, the zenith angle is 0° and the relative air mass is one that is written as AM1. When the Sun is at a zenith angle of 60°, the relative air mass is two, or AM2. At sunrise and sunset, the relative air mass is estimated to be ~38, or AM38 [17], as shown in Figure 2.5.

The condition represented by AM1 (i.e., $\theta = 0°$) requires either solar noon at the equator at equinox or solar noon at the Tropic of Cancer (Capricorn) at summer solstice. The minimum relative air mass, $AM_{(min)}$, at solar noon that may be achieved for a given latitude and declination angle is approximated by:

$$AM_{(min)} \approx [\cos(L - D)]^{-1}, \quad (2.10)$$

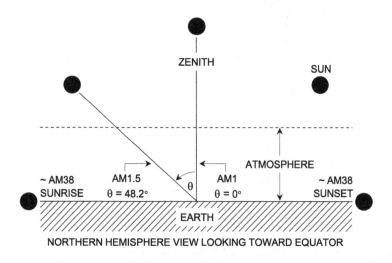

Fig. 2.5. Simplified schematic of optical air mass.

where L is latitude and D is declination angle. Equation 2.10 is a good approximation when $|L| < 66°$. For example, in Phoenix, AZ, USA ($L = \sim33.4°$ N), the minimum relative air mass at solar noon on the summer solstice ($D = 23.45°$) is $\sim$AM1.02 compared to $\sim$AM1.83 at solar noon on the winter solstice ($D = -23.45°$). The minimum relative air mass at solar noon on the summer solstice in Minneapolis, MN, USA ($L = \sim44.9°$ N) is $\sim$AM1.07 compared to $\sim$AM2.71 at solar noon on the winter solstice.

The solar photovoltaic community has decided that a terrestrial solar spectrum based on a relative air mass of 1.5, or AM1.5 (corresponding to $\theta \approx 48.2°$), represents a suitable reference spectrum with which to measure and compare the performance of different solar photovoltaic cells and modules as well as for the purpose of officially certifying new world record cells and modules [19]. The synthetic reference AM1.5 spectra (direct and global irradiance) were developed by Gueymard using his well-known modeling program, simple model of the atmospheric radiative transfer of sunshine (SMARTS) [20]. These spectra have been adopted by the American Society for Testing and Materials (ASTM) as the ASTM G173 reference standard [17].

2.12 AEROSOLS AND SCATTERING

Solar photons radiated from the Sun toward the Earth that actually reach all the way to the surface of the Earth are those that were neither absorbed nor reflected back into space by interaction with atmospheric constituents such as ozone (O_3), dust, volcanic ash, soot, smoke, smog, sea salts, water (H_2O) vapor and water droplets or ice crystals in clouds, oxygen (O_2), carbon dioxide (CO_2), methane (CH_4), and nitrogen dioxide (NO_2). Atmospheric H_2O and CO_2, for example, absorb certain wavelengths of IR radiation whereas O_3 is able to absorb much, but not all, of the incident UV radiation below 350 nm. If we consider a daily time period, then water vapor and ozone are more variable than carbon dioxide and methane [17].

Gas molecules may scatter solar photons in a process known as Rayleigh scattering [21]. From Earth, the sky appears blue during the bulk of the day because shorter wavelength photons (corresponding to violet and blue light) are scattered more effectively in Earth's atmosphere than longer wavelength photons. In other words, the intensity of scattering is governed by the following relationship:

$$I_{\text{Rayleigh}} \propto \lambda^{-4},\tag{2.11}$$

where λ is photon wavelength. Therefore, from Equation 2.11, it is evident that shorter wavelength violet and blue light will be scattered more than longer wavelength red light. There is also Mie scattering [22] that is caused when photons interact with water droplets (such as clouds) and particulates that are commonly known as aerosols (e.g., dust, volcanic ash, soot, smoke, and sea salts). Sunlight is also scattered by the ground, and this reflected light is known as albedo light. Albedo is the ratio of reflected to incident sunlight noting that albedo values range between 0 and 1 with, specifically, values of ~0.85 for fresh snow versus ~0.1 for dark surfaces [17].

Attenuation of photons by atmospheric aerosols may be characterized by the AOD, where the AOD may be approximated at any photon wavelength, as proposed by Ångström [23], with the following simple expression:

$$\text{AOD}(\lambda) = \beta\lambda^{-\alpha},\tag{2.12}$$

where β is the Ångström turbidity coefficient, λ is the photon wavelength (in μm), and α is the Ångström turbidity exponent. Clean air locations may have β = ~0.015 versus locations with heavy air pollution, dust, or smoke where β = ~0.5–2. Note that aerosols typically have values of α between ~0.5 and 2.5 [17]. It is worth pointing out that for the ASTM G173 standard, AOD = 0.084 (at λ = 500 nm), which is germane to the relatively "clean" air conditions that are predominant in the southwest USA [19]. Note here that most aerosols tend to scatter photons although smoke and soot tend to absorb photons. Another important parameter to consider for modeling terrestrial solar spectra is the precipitable water given by the height (in cm) of all the atmospheric water that would be condensed in a virtual vertical column. Precipitable water ranges from ~6 cm (when atmospheric conditions are described as being hot and humid) to ~0.2 cm (when atmospheric conditions are described as being cold and dry) [17].

2.13 CLOUDS

Clouds attenuate sunlight by scattering the light when the cloud cover is light to medium and by absorption when the cloud cover is heavy (dense). The optical depth of the clouds and also the fraction of the sky covered by clouds are important for understanding sunlight transmission. Dense (dark) clouds at low altitude have optical depth ranging from ~50 to 90. Thin and high-altitude clouds have optical depth ranging from ~1 to 5. If the clouds are sufficiently thick, then direct beam radiation is not transmitted to the ground; instead, it is the scattered (diffuse) indirect light that is available [17]. Direct and diffuse sunlight will be discussed next.

2.14 DIRECT VERSUS GLOBAL RADIATION

Sunlight that is not scattered is known as direct beam radiation. When the direct beam sunlight illuminates a plane that is normal to the direction of the Sun, this is known as direct normal irradiance or DNI. Therefore, any given terrestrial spectrum is actually composed of two primary constituents: direct and diffuse (or scattered) radiation. The direct plus diffuse radiation is known as global radiation. Therefore, the reference AM1.5 spectrum is delineated into the AM1.5G and AM1.5D spectra, where "G" represents global irradiance and "D" represents direct irradiance (plus a small circumsolar component).

For convenience, the synthetic reference AM1.5G spectrum (ASTM G173) has been developed with a total irradiance of $1000 \ W \ m^{-2}$ illuminating a ground-mounted collector (e.g., solar photovoltaic panel) that has a Sun-facing surface tilt of 37°. The specific tilt angle of 37° represents the average latitude in the contiguous USA, noting that for fixed-tilt solar photovoltaic panels, the ideal tilt angle may be set approximately equal to the site latitude to maximize annual energy yield. The AM1.5D spectrum (ASTM G173) has a total irradiance of $900 \ W \ m^{-2}$. The AM1.5 spectra span 280–4000 nm [17]. The AM1.5 spectra are shown in Figure 2.6. Direct solar radiation will be discussed more in Chapter 4.

The standard synthetic AM1.5 (ASTM G173) spectra were modeled and developed with SMARTS (v2.9.2) with the following assumptions: clear sky conditions (no clouds), AM0 irradiance = $1367 \ W \ m^{-2}$,

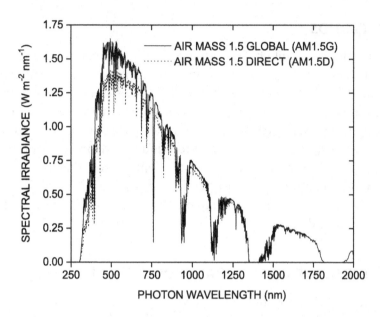

Fig. 2.6. ASTM G173 AM1.5 global and direct solar spectra (plotted out to 2000 nm) developed by Gueymard using the SMARTS modeling code [19]. Water vapor has pronounced absorption in the IR from 920 to 980 nm, 1100 to 1200 nm, 1300 to 1500 nm, and 1750 to 1950 nm, whereas ozone absorbs strongly below 350 nm (hence, attenuation of UV light) and also weakly in the visible [17].

Sun-facing surface tilt (e.g., of a ground mounted solar collector) = 37°, site temperature = 288.1 K, site pressure = 1013.25 mbar, ground altitude = 0 km, relative humidity = 46.04%, precipitable water = 1.416 cm, ozone = 0.3438 atm-cm, rural aerosol model with AOD = 0.084 at λ = 500 nm, CO_2 volumetric mixing ratio = 370 ppm, spectral albedo based on "light sandy soil" with non-Lambertian reflection, and a pyrheliometer opening half angle = 2.9° (for the purpose of calculating circumsolar radiation).

2.15 PHOTON FLUX

Photon flux will become important later in Chapter 3 when calculating photogenerated current density in solar photovoltaic cells. Under a clear sky AM1.5G terrestrial spectrum, there are some 10^{17} photons irradiating a 1 cm² Sun-facing surface area every second. Spectral photon flux $\varphi(\lambda)$ is calculated from the spectral irradiance I_λ (W m⁻² nm⁻¹) through the following relationship:

$$\varphi(\lambda) = I_\lambda / E, \qquad (2.13)$$

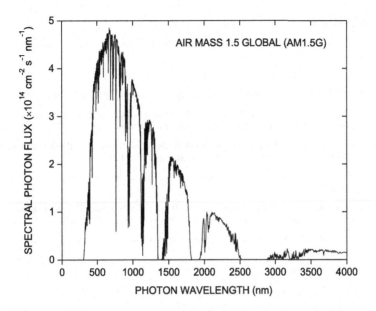

Fig. 2.7. ASTM G173 AM1.5G spectral photon flux (plotted out to the full 4000 nm cutoff).

where $E = hc/\lambda$ and h is Planck's constant. The AM1.5G spectral photon flux is shown in Figure 2.7, and the integrated photon flux is shown in Figure 2.8.

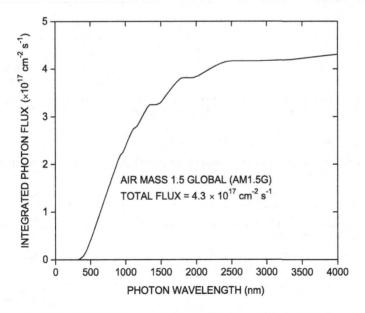

Fig. 2.8. ASTM G173 AM1.5G integrated photon flux (plotted out to the full 4000 nm cutoff).

REFERENCES

[1]   K.R. Lang, The Cambridge Encyclopedia of the Sun, Cambridge University Press, Cambridge, (2001).

[2]   B. Buffett, Earth's enigmatic inner core, Physics Today 66 (2013) 37–41.

[3]   J.N. Bahcall, Neutrino Astrophysics, Cambridge University Press, New York, (1989).

[4]   E.G. Adelberger, A. García, R.G. Hamish Robertson, K.A. Snover, et al. Solar fusion cross sections II: the pp chain and CNO cycles, Reviews of Modern Physics 83 (2011) 195.

[5]   G. Bellini, J. Benziger, D. Bick, G. Bonfini, D. Bravo, B. Caccianiga, et al. Neutrinos from the primary proton–proton fusion process in the Sun, Nature 512 (2014) 383–386.

[6]   H.A. Bethe, C.L. Critchfield, The formation of deuterons by proton combination, Physical Review 54 (1938) 248.

[7]   G. Bellini, J. Benziger, D. Bick, S. Bonetti, G. Bonfini, D. Bravo, et al. First evidence of *pep* solar neutrinos by direct detection in Borexino, Physical Review Letters 108 (2012) 051302.

[8]   Super-Kamiokande Detector. Available from: http://www-sk.icrr.u-tokyo.ac.jp/index-e.html.

[9]   P.A. Čerenkov, Visible emission of clean liquids by action of γ radiation, Doklady Akademii Nauk SSSR 2 (1934) 451.

[10] G. Gamow, Zur quantentheorie des atomkernes, Zeitschrift für Physik 51 (1928) 204–212.

[11] USGS Mineral Commodity Summaries. Available from: http://minerals.usgs.gov/minerals/pubs/mcs/.

[12] National Ignition Facility. Available from: https://lasers.llnl.gov/.

[13] International Thermonuclear Experimental Reactor. Available from: http://www.iter.org/.

[14] Laboratory for Atmospheric and Space Physics. Available from: http://lasp.colorado.edu/home/.

[15] G. Kopp, G. Lawrence, The total irradiance monitor (TIM): instrument design, Solar Physics 230 (2005) 91–109.

[16] C.A. Gueymard, The sun's total and spectral irradiance for solar energy applications and solar radiation models, Solar Energy 76 (2004) 423–453.

[17] C.A. Gueymard, D. Myers, Solar resources for space and terrestrial applications, in: L. Fraas, L. Partain (Eds.), Solar Cells and their Applications, John Wiley & Sons, Hoboken, (2010).

[18] C.A. Gueymard, Direct solar transmittance and irradiance predictions with broadband models. Part I: detailed theoretical performance assessment, Solar Energy 74 (2003) 355–379.

[19] C.A. Gueymard, D. Myers, K. Emery, Proposed reference irradiance spectra for solar energy systems testing, Solar Energy 73 (2002) 443–467.

[20] C.A. Gueymard, Parameterized transmittance model for direct beam and circumsolar spectral irradiance, Solar Energy 71 (2001) 325–346.

[21] J.W. Strutt, XV. On the light from the sky, its polarization and colour, The London, Edinburgh, and Dublin Philosophical Magazine and Journal of Science 41 (1871) 107–120.

[22] G. Mie, Beiträge zur Optik trüber Medien, speziell kolloidaler Metallösungen, Annalen der Physik 330 (1908) 377–445.

[23] A. Ångström, On the atmospheric transmission of sun radiation and on dust in the air, Geografiska Annaler 11 (1929) 156–166.

CHAPTER 3

Device Operation

3.1 INTRODUCTION

This chapter begins with a brief history of some of the direct development and deployment of solar photovoltaic cells or scientific discoveries and theories that ultimately paved the way for a more complete understanding of solar photovoltaic cells including ways to improve their fabrication and performance. Next, $p-n$ junction solar photovoltaic cells are introduced followed by a look at the relationship between bandgap energy and power conversion efficiency. Then, the photogenerated current density, and its relationship to solar photon flux, is presented. Next, absorption coefficient is discussed. An explanation of hot-carrier relaxation is offered, which then leads to the open circuit voltage and its relation to chemical potential and quasi-Fermi levels. The principle of detailed balance is then invoked to actually quantify the open circuit voltage for solar photovoltaic cells with backside mirrors, air-exposed top and bottom surfaces (bifacial design), and backside parasitically absorbing substrates. Following this, the necessary steps required to calculate power conversion efficiency are introduced. Then, there is a discussion about free energy management in solar photovoltaic cells. Next, the radiative recombination coefficient and radiative lifetime is explained, followed by a discussion on nonradiative Auger and Shockley–Read–Hall (SRH) recombination. The role of minority carrier diffusion length is then introduced. The discussion shifts to multijunction solar photovoltaic cells with examples of monolithic, series-connected devices that exhibit greater power conversion efficiency than standard single junction solar photovoltaic cells. Then, there is a discussion on the challenges underlying successful hot-carrier solar photovoltaic cell operation. Finally, there is a discussion on device engineering details required to optimize real solar photovoltaic cells. The goal of this chapter is to provide a solid foundation for understanding the basic physics of solar photovoltaic cells and their operation.

Solar Photovoltaic Cells: Photons to Electricity. http://dx.doi.org/10.1016/B978-0-12-802329-7.00003-1

3.2 HISTORY

The field of photovoltaics is 175 years old as of the year 2014. Some, but certainly not all, of the important and interesting developments in this field (or related fields) are outlined next.

In 1839, the 19-year old French scientist A. E. Becquerel discovered the photovoltaic effect (known previously as the Becquerel effect and perhaps also stated here as a photoelectrochemical effect) when metal halide-coated platinum (Pt) electrodes immersed in an aqueous solution were exposed to light [1,2]. Then, in 1873, W. Smith reported photoconductivity in bars of selenium (Se) [3]. In 1874, F. Braun discovered the rectifier effect in metal sulfides with metallic point contacts (published in 1875) [4]. In 1875 through 1876, in an epic treatise, J. W. Gibbs presented the concept of chemical potential [5]. The voltage of a solar photovoltaic cell is related to the difference in chemical potential of the photogenerated and relaxed electrons and holes. During his PhD work in 1879, E. Hall discovered the Hall effect [6], which is now used routinely to characterize semiconductor carrier type and concentration. By 1887, H. Hertz had discovered the photoelectric effect in metal electrodes exposed to ultraviolet (UV) light [7].

In 1900, M. Planck developed a theory to correctly explain black body radiation (published in 1901) [8], and in doing so, established (somewhat reluctantly at first) the relationship between photon energy E and resonator frequency v expressed as $E = hv$, where h is known as Planck's constant. Soon after, in 1905, A. Einstein realized that photons are quanta of energy and used this to develop a theory [9] to explain the photoelectric effect observed earlier by Hertz in 1887. Therefore, it is often considered that Planck and Einstein helped to commence the era of quantum mechanics. In 1916, the Polish scientist J. Czochralski discovered a method (published in 1918) [10] to grow single crystals now known eponymously as the Czochralski process – a technique still used today to produce single crystal boules of semiconductors such as silicon (Si) that are sliced into wafers, which may then be processed into solar photovoltaic cells.

By 1933, E. Wigner and F. Seitz had developed a band theory of sodium metal, which was more realistic than earlier attempts to model electron bands [11]. Subsequent developments in the field allowed semiconductors to be accurately modeled. R. Ohl, an engineer trained in electrochemistry working at Bell Telephone Laboratories, and coworkers had discovered p–n junctions in 1939 on into 1940 in purified Si that had trace amounts of impurities. In 1940, Ohl observed the photovoltaic effect in his Si samples when they were illuminated by a 40 W incandescent desk lamp [12]. These p–n junctions were known originally as "barriers" and are now used routinely in solid-state devices such as high-efficiency solar photovoltaic cells. By 1941, Ohl had filed a patent concerning Si solar photovoltaic cells [13]. Silicon solar photovoltaic cells (single crystal, multicrystal, and amorphous variants) dominate the market as of 2014. In 1947, J. Bardeen discussed the physics of surface states in semiconductors [14]; the controlled passivation of surface states is important for solar photovoltaic cells (which are minority carrier devices) to mitigate undesirable surface recombination. By 1949, W. Shockley had developed a theory of semiconductor p–n junctions [15]. In 1951 into 1952, R. Hall [16,17], and also in 1952, W. Shockley and W. Read, Jr. [18] offered explanations of nonradiative recombination in semiconductors, now known as SRH recombination for SRH recombination. The goal in solar photovoltaic cell design and manufacturing is to reduce SRH recombination; an ideal solar photovoltaic cell at open circuit would be radiatively limited (i.e., dominated by radiative recombination).

In 1953, at a symposium in Madison, Wisconsin, P. Flinn and D. Trivich mentioned the concept of multiple junction solar photovoltaic cells (published in 1955) [19]. Also in 1953, P. Rappaport at RCA discussed the electron–voltaic effect in Si and germanium (Ge) p–n junctions exposed to beta particles from a ^{90}Sr–^{90}Y radioactive source (published in 1954) [20]. In 1954, D. Chapin, C. Fuller, and G. Pearson fabricated more efficient Si solar photovoltaic cells [21] (also known then by the term "solar battery") that were widely heralded by the management at Bell Labs. And in 1954, D. Reynolds and coworkers reported the photovoltaic effect in cadmium sulfide (CdS) [22]. Also in 1954, W. van Roosbroeck and W. Shockley published a paper regarding the use of the principle of detailed balance to calculate the radiative recombination of electrons and holes in Ge [23]. D. Jenny, J. Loferski, and P. Rappaport developed functional diffused-junction gallium arsenide (GaAs) solar photovoltaic cells in 1955 (published in 1956) [24]. Also in 1955, E. Jackson from Texas Instruments discussed the benefits of multijunction solar photovoltaic cells [25] and filed a detailed patent regarding multijunction solar photovoltaic cells [26]. In 1957, C.-T. Sah, R. Noyce, and W. Shockley published an improved theory of p–n junctions [27]. Also in 1957, L. Esaki discovered p–n tunnel (or Esaki) diodes in Ge (published in 1958) [28]. Tunnel diodes are used in monolithic multijunction solar photovoltaic cells to electrically, mechanically, and optically connect individual subcells. In 1958, the USA launched the 1.47 kg Vanguard I satellite, managed by the Naval Research Laboratory (NRL). This was the first satellite powered with Si solar photovoltaic cells and is the oldest manmade object still in space [29]. Also in 1958, D. Kearns and M. Calvin observed the photovoltaic effect in organic junctions (magnesium phthalocyanine/air-oxidized tetramethyl p-phenylenediamine) [30]. Then, in 1959, Rappaport reported data on functional solar photovoltaic cells comprised of cadmium telluride (CdTe) and indium phosphide (InP) [31]. In 1960, R. Anderson reported epitaxially grown (and approximately lattice matched) single crystal Ge–GaAs heterojunctions [32].

In what is now considered a seminal paper, although the original manuscript had been rejected by the *Journal of Applied Physics* [33], W. Shockley and H. Queisser calculated the limiting efficiency of single junction solar photovoltaic cells in 1960 by employing the principle of detailed balance as published in 1961 [34]. In 1963, Z. Alferov and R. Kazarinov [35] and independently H. Kroemer [36] discussed the concept of double heterostructures (DH) for the purpose of vastly improving semiconductor diode injection lasers; this same concept is used today in high-efficiency III–V solar cells where wider bandgap window and back surface field layers are utilized to clad a narrower bandgap absorber to provide surface passivation as well as minority carrier blocking. H. Gerischer and coworkers reported organic molecule dye-sensitized photocurrent in the inorganic semiconductor n-type zinc oxide (ZnO) and the organic semiconductor p-type perylene in 1967 (published in 1968) [37]. In 1970, Z. Alferov and coworkers reported AlGaAs–GaAs heterojunction solar cells [38]. K. Böer led the development of a practical house at the University of Delaware in 1973 known as Solar One that used a hybrid system of solar photovoltaic cells for electricity and solar thermal collectors for heat [39]. S. Wagner and coworkers at Bell Labs reported CdS/CuInSe$_2$ heterojunction photovoltaic detectors in 1974 [40]. In 1975, C. Backus discussed the benefits of sunlight concentration (up to 1000×) for both solar photovoltaic and solar thermal systems [41]. In 1976, D. Carlson and C. Wronski at RCA developed thin amorphous silicon (a-Si) solar photovoltaic cells [42]. S. Bedair, M. Lamorte, and J. Hauser reported a functional, monolithic AlGaAs/GaAs double junction solar photovoltaic cell with an integral tunnel diode in 1978 (published in 1979) [43]. In 1980, C. Henry at Bell Labs published a thorough detailed balance analysis of multijunction solar photovoltaic cell efficiency (including 1000× sunlight concentration using the terrestrial AM1.5 spectrum) [44].

By 2013, M. Taguchi and coworkers at Panasonic had developed a practical Si solar photovoltaic cell (101.8 cm^2 area, 98 μm thick n-type Czochralski single crystal wafer) with a power conversion efficiency of 24.7% (AM1.5G spectrum at 298 K) and an open circuit voltage of 0.75 V. This cell was constructed with thin (~10 nm) surface passivation layers (top and bottom) of undoped amorphous Si followed by thin (~10 nm) layers of doped amorphous Si in an architecture known as HIT® – heterojunction with intrinsic thin layer (published in 2014) [45]. Note that Panasonic later reported a 25.6% efficient Si cell in 2014 (see "technical content note" at the end of the book for more details). As of 2013, the best solar photovoltaic module (1092 cm^2 aperture area, fabricated by Amonix) was based on III–V triple junction cells operating under concentrated sunlight through the use of Fresnel lenses and reported to have reached 35.9% module efficiency at 298 K cell temperature and 1000 W m^{-2} direct irradiance [46].

From this brief historical review, we see that by ~1980, the primary groundwork was in place for understanding and developing high-efficiency solar photovoltaic cells. Between 1980 and the date of this book (2014), the advancements in solar photovoltaic cells have been steadily evolutionary as expected for a maturing technology that had initial beginnings 175 years ago. It is therefore thoroughly expected that solar photovoltaic cell (and module) efficiency will continue to see evolutionary improvement. Nonetheless, there are still many exciting and yet entirely pragmatic developments to be made, such as with single and multiple junction solar photovoltaic cells operating under sunlight concentration, bifacial solar photovoltaic cell designs, reduction of semiconductor material and the pursuit of thinned cells, innovative module design and deployment methodologies, scalability and integration, balance of systems improvements, installation cost reduction, and so forth.

Meaningful new developments in the field of solar photovoltaics do not require "3rd Generation" technology that typically means solar photovoltaic cells based on concepts such as hot-carrier extraction, intermediate bands, or multiple electron-hole pair generation per absorbed photon [47]. In other words, while these concepts may be interesting (at least esoterically speaking), the point is that they actually do not represent a critical or bottlenecking technology gap that we must somehow wait to overcome to realize improvements in solar photovoltaic cell and module efficiency or more substantial and widespread utilization of solar electricity. Therefore, in this book, there will be limited discussion on these cell concepts with the exception of a brief discussion on hot-carrier solar photovoltaic cells. Intermediate band solar photovoltaic cells appear to only offer the same limiting power conversion efficiency as conventional 3-junction cells while the purported benefits of the multiple electron-hole pair generation concept appear so far to be questionable [47]. Furthermore, we will see later in Chapter 4 that 6-junction solar photovoltaic cells operating under sunlight concentration offer a pragmatic path to both high-efficiency cells and modules, and a way to compete directly with, or at least help supplement, the most efficient thermal power plants based on combined cycle natural gas and steam turbines for electricity generation. Note that multijunction solar photovoltaic cells should not be conflated with "3rd Generation" concepts since multiple junctions were already proposed as early as the mid 1950s and then successfully fabricated beginning in the late 1970s as outlined earlier.

3.3 P–N JUNCTION CELLS

The solar photovoltaic cell is a beautiful and elegant device that absorbs solar radiation (photons) and internally converts the solar radiant energy ultimately into direct current (DC) electricity without acoustic or chemical pollution during operation. For now, we will sketch the primary constituent design attributes that comprise a p–n junction Si solar cell as shown in Figure 3.1.

The simplistic Si solar photovoltaic cell shown in Figure 3.1 is only one variant. In fact, this particular cell is a thoroughly generic and suboptimal design. Other variants include the a–Si:H/c–Si/a–Si:H DH design (using n-type c–Si wafers) such as that embodied by Panasonic's

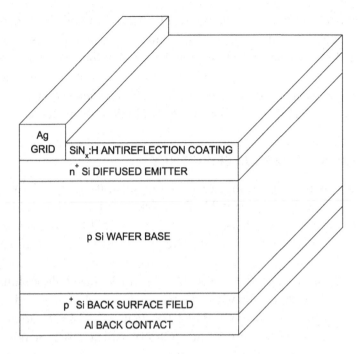

Fig. 3.1. Schematic of a generic Si solar photovoltaic cell. Si cells typically have a textured front surface (not shown here) to aid light coupling due to the weak absorption characteristic of indirect bandgap Si. This view only shows a single grid line for clarity. (After Ref. [48].)

HIT® (and HIT® Double bifacial) cells and the all-back contact design (using n-type c–Si wafers) as embodied by SunPower's Maxeon® cell. In other words, there is no single, fixed way to fabricate a functional solar photovoltaic cell even within the same semiconductor material such as crystalline Si. In fact, one common myth is that a traditional p–n junction is required as shown in Figure 3.1. However, there are other designs including Schottky barrier, metal–insulator–semiconductor (MIS), and dye-sensitized solar photovoltaic cells that do not actually require a traditional p–n junction. Nonetheless, p–n and p–i–n junction cells do offer high efficiency and stability and so they will represent much of the focus here.

Since p–n junction theory and its application specifically to devices such as solar photovoltaic cells has already been thoroughly addressed in well-known texts such as *Electrons and Holes in Semiconductors* by Shockley [49] and *Physics of Semiconductor Devices* by Sze [50], there will not be an extensive repetition of this material here except when necessary to highlight specific points of interest. In addition, there will not be a repetition of semiconductor band theory as this has also been thoroughly addressed in texts such as *Semiconductors: Bonds and Bands* by Ferry [51]. Finally, more subtle details about statistical and thermal physics is adequately discussed in texts such as *Elementary Statistical Physics* by Kittel [52] and *Thermal Physics* by Kittel and Kroemer [53].

3.4 BANDGAP VERSUS EFFICIENCY

Silicon, with an energy bandgap at room temperature of 1.12 eV, is not the only semiconductor suitable for efficient solar photovoltaic cells. By invoking the principle of detailed balance [52] (to be discussed in Section 3.9), a plot of theoretical power conversion efficiency versus cell absorber bandgap energy (AM1.5G spectrum, T_{cell} = 298 K) is shown in Figure 3.2. From Figure 3.2, the peak radiative (detailed balance limiting) efficiency of 33.7% results when the bandgap is 1.34 eV; coincidentally, the semiconductor InP has a direct bandgap of 1.34 eV. Other ideal semiconductors are $CuIn_xGa_{1-x}Se_2$ (CIGS) with a direct bandgap of 1.15 eV and peak radiative efficiency of 33.5%, Si with an indirect

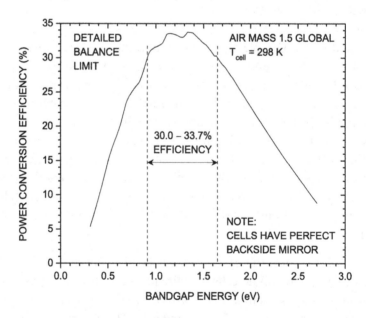

Fig. 3.2. Ideal power conversion efficiency versus bandgap energy. Peak efficiency corresponds to bandgaps from 0.91–1.65 eV. (After Ref. [48].)

bandgap of 1.12 eV and peak radiative efficiency of 33.4% (although Si is thought to be fundamentally limited instead by nonradiative Auger recombination), GaAs with a direct bandgap of 1.42 eV and peak radiative efficiency of 33.2%, and CdTe with a direct bandgap of 1.50 eV and peak radiative efficiency of 32.1%. Taking into consideration single junction solar photovoltaic cells, as of 2014 for small area laboratory devices, only GaAs cells (28.8% record efficiency) [46] have so far exceeded the performance of Si cells (25.6% record efficiency). Silicon is abundant in the Earth's crust as it is sourced from silicon dioxide (SiO_2), relatively low-cost, chemically stable, and easily doped n-type and p-type. The indirect bandgap is not considered ideal, yet Si is still a good choice for solar photovoltaic cells. As of 2014, single junction GaAs solar photovoltaic cells are not generating any meaningful electricity as the market is completely dominated by Si with a small fraction supplied by polycrystalline thin-film CdTe and an even smaller fraction supplied by polycrystalline thin-film CIGS. No meaningful amount of electricity was being generated on Earth as of 2014 with InP, organic molecule (including polymer), dye-sensitized, perovskite, $Cu_2ZnSnS_xSe_{4-x}$ (CZTSS), or any of the so-called "3rd Generation" cells.

3.5 PHOTOGENERATED CURRENT DENSITY

Solar photovoltaic cells are threshold energy conversion devices [54] because they have a discrete energy bandgap E_g. The photogenerated current density J_{ph} of a solar photovoltaic cell is a function of E_g. Note that photon wavelength and energy are related by:

$$E = hv = hc / \lambda, \tag{3.1}$$

where v is photon frequency. Alternatively, the photon energy may also be written as $E = \hbar\omega$ where $\hbar$ is reduced Planck's constant ($\hbar = h / 2\pi$) and ω is angular frequency. Photons with energy $hv \geq E_g$ may be absorbed in an interband (valence band to conduction band) process with adequate thickness of the semiconductor, which is primarily a function of the absorption coefficient of the semiconductor. At first order, photons with energy $hv < E_g$ are not absorbed and therefore do not contribute to the photogenerated current density. Photogenerated current density for solar photovoltaic cells under AM1.5G illumination is shown in Figure 3.3 where it is assumed that each absorbed photon

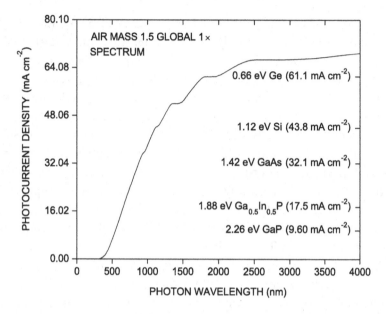

Fig. 3.3. Photogenerated current density including maximum current density for select semiconductors (AM1.5G spectrum). The tick marks on the right-hand side represent the photocurrent density for the selected semiconductors shown here. (After Ref. [48].)

generates one electron and one hole. The integrated solar photon flux Φ, shown previously in Figure 2.8 (Chapter 2), is multiplied by the electronic charge e to calculate the photogenerated current density:

$$J_{ph} = e \cdot \Phi. \tag{3.2}$$

3.6 ABSORPTION COEFFICIENT

A useful way to characterize photon absorption, discussed in Section 3.5, is through the absorption coefficient $\alpha(E)$, which, for direct interband transitions, is expressed as:

$$\alpha(E) = \frac{\pi e^2}{c\varepsilon_0 n_r m_e^2 \omega} \int \left| \langle f | \mathbf{a} \cdot \mathbf{p} | i \rangle \right|^2 g_i(E_i) g_f(E - E_g - E_i) dE_i, \tag{3.3}$$

where c is velocity of photons in vacuum, ε_0 is vacuum permittivity, n_r is semiconductor refractive index, m_e is electron mass, ω is photon angular frequency, $|i\rangle$ and $\langle f|$ are initial and final states respectively, $\mathbf{a}$ is polarization vector, $\mathbf{p}$ is momentum operator, $\int g_i(E_i) g_f(E - E_g - E_i) dE_i$ is joint electronic density of states, and E is photon energy. A derivation of the absorption coefficient is included in Appendix E.

Overall, it is important to know the absorption coefficient to select the proper thickness of the base layer in a solar photovoltaic cell so that precious semiconductor material is not unduly wasted. In an indirect bandgap semiconductor such as Si, optical transitions are indirect and therefore require absorption (or emission) of a phonon to account for the difference in wave vector between the initial (valence) and final (conduction) band states. To be clear, Si does have a wide direct gap estimated at ~4.2 eV, yet there are few solar photons with energy greater than 4.2 eV in the terrestrial solar spectrum. A plot of the absorption coefficient of direct bandgap GaAs, from experimental data tabulated in Adachi's *Optical Constants of Crystalline and Amorphous Semiconductors* book [55], is shown in Figure 3.4.

3.7 HOT-CARRIER RELAXATION

The process of carrier relaxation refers to the condition of the above bandgap photogenerated electrons and holes (i.e., hot carriers) interacting with lattice phonons, therefore, converting their excess kinetic

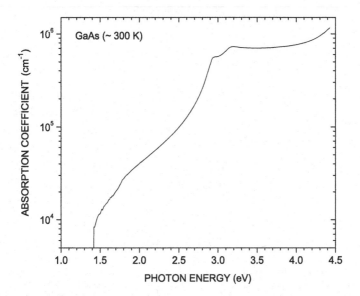

Fig. 3.4. Absorption coefficient of direct bandgap GaAs. Experimental source data is from Ref. [55].

energy into thermal energy until they reach to/near their fundamental band edges – this entropy generating relaxation process represents a loss in solar photovoltaic cells as shown in Figure 3.5. When the carriers are initially photogenerated, they continuously experience carrier–carrier scattering that results in a "thermal" population of carriers described by $T_{\text{hot carrier}} > T_{\text{lattice}}$. The hot carriers then incrementally lose their excess (with respect to the bandgap) energy as they interact with lattice phonons, and this process is known as "relaxation". For example, in GaAs, a hot electron may interact multiple times with longitudinal optical (LO) phonons with each interaction resulting in the electron discretely losing ~36 meV energy as it scatters in the conduction band from one quantum state to a slightly lower energy quantum state. The carrier relaxation process is rapid, with the duration of each electron–phonon collision typically occurring in the subpicosecond timescale and complete carrier relaxation to/near the fundamental band edges typically occurring within the few picosecond timescale. Electrons may not necessarily relax all the way to the bottom of the conduction band (Γ valley in GaAs) if their remaining excess energy with respect to the bottom of the conduction band is less than the LO phonon energy (~36 meV). As a side

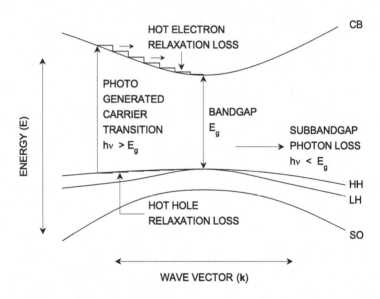

Fig. 3.5. Schematic of photogenerated carrier transition, hot-carrier relaxation, and subbandgap photon loss depicted near the Γ valley in direct bandgap GaAs. These bands were generated by following the technique in Ref. [57]. Key: CB is (lowest) conduction band, HH is heavy hole valence band, LH is light hole valence band, and SO is split-off hole valence band. (After Ref. [58].)

note, and again considering the specific case of GaAs, holes primarily interact with transverse optical (TO) phonons, but the overall relaxation mechanics are the same as for electrons. Typically, holes relax even faster than electrons due to the larger electronic density of states in the valence bands. Once the carriers have relaxed down to/near their respective band edges (and assuming they are not at too large a concentration), they are described to first order as being approximately "thermal", this time with $T_{\text{relaxed carrier}} \approx T_{\text{lattice}}$. This discussion provides a generic overview of carrier relaxation. The type of semiconductor (and specific band structure artifacts such as location of satellite valleys) as well as the concentration of the photoexcited carriers play a distinct role in carrier relaxation physics. A complete understanding of carrier relaxation requires in-depth modeling on a case-by-case basis. More details on carrier relaxation and transport may be found in Ferry's *Semiconductor Transport* book [56].

In good quality *p–n* junction solar photovoltaic cells, the hot-carrier relaxation timescale is usually multiple orders of magnitude less than the radiative and nonradiative recombination timescale (i.e., lifetime) of the relaxed carriers. Once relaxed, the approximately band-edge energy mobile carriers diffuse (and drift when in the presence of an internal electric field). Collectively, this transport behavior is known as drift–diffusion. Drift–diffusion physics is routinely used in software codes for the purpose of modeling devices such as bulk *p–n* junction solar photovoltaic cells. Later, the helpful concept of minority carrier (i.e., electrons in a *p*-type layer or holes in an *n*-type layer) diffusion length will be presented in Section 3.14. It is worth noting that drift–diffusion (i.e., semiclassical) physics is not applicable to situations where quantum transport physics may be considered, such as transport through degenerately doped tunnel (Esaki) diodes that are used to interconnect individual subcells in two-terminal, series-connected multijunction solar photovoltaic cells. In addition, standard drift–diffusion physics is not applicable for modeling hot-carrier transport where $T_{\text{hot carrier}} > T_{\text{lattice}}$.

3.8 OPEN CIRCUIT VOLTAGE

Under steady-state illumination, the open circuit voltage V_{oc} of a solar photovoltaic cell is related to the difference in quasi-Fermi levels $(E_{F_n} - E_{F_p})$, which is also known as the difference in chemical potential

$\Delta\mu$ between the photogenerated electrons and holes that have relaxed to their fundamental band edges (approximately the bottom of the conduction band for electrons and top of the valence band for holes). In other words:

$$eV_{oc} = \Delta\mu = (E_{F_n} - E_{F_p}). \tag{3.4}$$

A light-bias band diagram is a good way to visualize the relationship between quasi-Fermi levels and voltage. As an example, a light-bias band diagram of a 1.42 eV GaAs solar photovoltaic cell that has 1.55 eV $Al_{0.10}Ga_{0.90}As$ window and back surface field (BSF) layers is shown in Figure 3.6.

3.9 DETAILED BALANCE

Now, if we wish to know the maximum V_{oc} that may be obtained by an ideal solar photovoltaic cell, we can employ the principle of detailed balance as originally discussed by Shockley and Queisser [34]. Within the standard Boltzmann approximation and also by assuming constant quasi-Fermi level separation under steady-state sunlight illumination as well as the absence of any parasitic resistive power loss ($P_{loss} = I^2R$ where

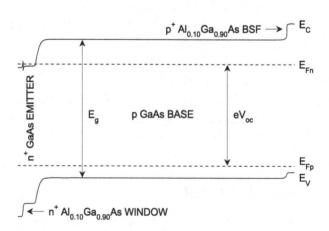

Fig. 3.6. Sunlight-biased band diagram of a GaAs solar photovoltaic cell with thin GaAs emitter as well as thin $Al_{0.10}Ga_{0.90}As$ window and back surface field (BSF) layers generated by using PC1D [59]. (After Ref. [58].)

I is current and R is resistance), the maximum V_{oc} occurs when the only sink for a photogenerated (and subsequently relaxed) electron at/near the bottom of the conduction band is through an interband radiative recombination process with a valence band (i.e., ground state) hole whereby a luminescent photon is emitted. Therefore, the detailed balance limit is also known as the radiative limit. This detailed balance approach may be used to investigate the maximum V_{oc} for solar photovoltaic cells [60] with (*Case A*) a backside mirror, (*Case B*) an air-exposed backside, which is also known as a bifacial configuration, and (*Case C*) a parasitically absorbing backside substrate.

(*Case A*) The detailed balance-limiting V_{oc} for a solar photovoltaic cell with a perfect backside mirror is given by:

$$V_{oc} = E_g e^{-1} - k_B T e^{-1} \ln(2\pi e E_g^2 k_B T h^{-3} c^{-2} J_{ph}^{-1}), \qquad (3.5)$$

where k_B is Boltzmann's constant and T is cell temperature.

(*Case B*) The detailed balance-limiting V_{oc} for a solar photovoltaic cell with an air-exposed backside is given by:

$$V_{oc} = E_g e^{-1} - k_B T e^{-1} \ln[2\pi(2) e E_g^2 k_B T h^{-3} c^{-2} J_{ph}^{-1}]. \qquad (3.6)$$

(*Case C*) The detailed balance-limiting V_{oc} for a solar photovoltaic cell with a parasitically absorbing backside substrate is given by:

$$V_{oc} = E_g e^{-1} - k_B T e^{-1} \ln[2\pi(1 + n_r^2) e E_g^2 k_B T h^{-3} c^{-2} J_{ph}^{-1}], \qquad (3.7)$$

where n_r is the refractive index of the parasitically absorbing substrate [61].

By comparing these three cases, we see that mirror cells have the largest V_{oc} whereas parasitic substrate cells have the smallest V_{oc}. Compared to an ideal mirror cell, the parasitically absorbing substrate cell has V_{oc} reduced by the factor of $-k_B T e^{-1} \ln(1 + n_r^2)$. This reduction in V_{oc} is due to the increase in entropy associated with luminescent radiative absorption in the parasitic backside substrate. A comparison of the detailed balance-limiting V_{oc} for direct bandgap InP, GaAs, CdTe, and $Ga_{0.5}In_{0.5}P$ solar photovoltaic cells with ideal backside mirrors versus parasitically absorbing backside substrates [61] is shown in Figure 3.7. A derivation of the detailed balance-limiting V_{oc} is included in Appendix F.

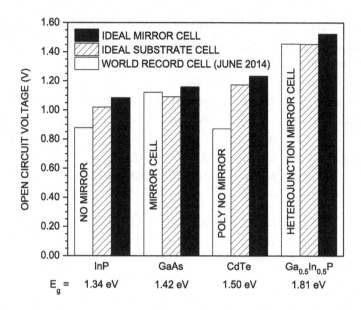

Fig. 3.7. Comparison of detailed balance-limiting open circuit voltage for ideal InP, GaAs, CdTe, and $Ga_{0.5}In_{0.5}P$ mirror and substrate solar photovoltaic cells versus reported world record cells as of June 2014, AM1.5G spectrum, 298 K. (After Ref. [61].)

Note here that due to momentum randomization, luminescent radiation may be emitted in any direction, which means on average that half of the luminescent radiation is emitted toward the backside of the cell. In a mirror cell, the luminescent radiation that is emitted toward the rear may be reflected toward the front. Then, within the framework of Snell's law, the luminescent radiation either escapes or is totally internally reflected. Luminescent photons that are internally reflected may be absorbed in the cell. Luminescent radiation emission and absorption is a normal process that has been known since at least 1957 [62], and, by default, it has already been included in the detailed balance V_{oc} discussed here.

3.10 POWER CONVERSION EFFICIENCY

Previously, we have focused attention on the open circuit condition. To generate the maximum power, solar photovoltaic cells operate at what is known as the maximum power point, as shown in Figure 3.8, where, as an approximation, J_{ph} has been replaced with the short circuit current density J_{sc}, which has become the standard nomenclature of interest

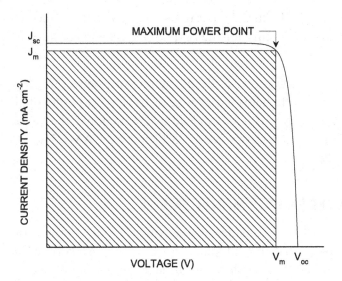

Fig. 3.8. Current density versus voltage curve showing the maximum power point. This plot was generated for a GaAs solar photovoltaic cell by using PC1D v5.9 [59].

in light J–V characterization (typically done indoors with a calibrated lamp and probe station).

Having already calculated J_{ph} and V_{oc}, we can then proceed to calculate the actual power conversion efficiency of a solar photovoltaic cell. To do this, we need to first set $d(JV)/dV = 0$ to find the maximum power point voltage V_m as given by:

$$V_m = V_{oc} - k_B T e^{-1} \ln(1 + e V_m / k_B T). \tag{3.8}$$

Rather than using numerical iteration to solve Equation 3.8, the following approximation holds:

$$V_m \approx V_{oc} - k_B T e^{-1} \ln(1 + e V_{oc} / k_B T). \tag{3.9}$$

From this, the maximum power point current density J_m is given by:

$$J_m = J_{ph} / (1 + k_B T / e V_m). \tag{3.10}$$

Finally, the power conversion efficiency η of the solar photovoltaic cell is then given by:

$$\eta = P_{out} / P_{in}, \tag{3.11}$$

where $P_{out} = J_m V_m$ and P_{in} is the incident solar irradiance (which, for example, is 0.1 W cm^{-2} for the AM1.5G spectrum at one Sun illumination) [61].

Often, a metric known as the fill factor, expressed as $FF = J_m V_m / J_{sc} V_{oc}$, is used as a way to help characterize or track solar photovoltaic cell performance. Series resistance can lead to a reduction in the fill factor. Note that an ideal GaAs mirror cell at 25°C under the AM1.5G spectrum has $FF \sim 0.89$.

3.11 FREE ENERGY MANAGEMENT

While we are still considering the topic of open circuit voltage, it is worth noting that the difference in free energy ΔF between the photo-generated and relaxed electrons at/near the bottom of the conduction band and holes at/near the top of the valence band is equivalent to the open circuit voltage multiplied by the electronic charge, or $\Delta F = eV_{oc}$. A figure of merit defined as $\Delta F / E_g$ may be invoked to understand the relevance of photon quality as shown in Figure 3.9. In other words,

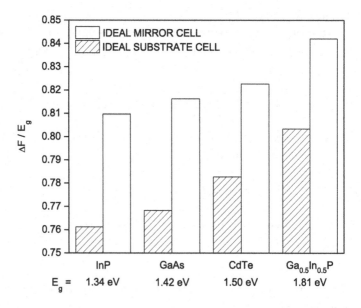

Fig. 3.9. Detailed balance-limiting figure of merit or change in free energy divided by the bandgap energy for ideal InP, GaAs, CdTe, and Ga$_{0.5}$In$_{0.5}$P mirror versus substrate solar photovoltaic cells, AM1.5G spectrum, 298 K. (After Ref. [61].)

solar photovoltaic cells with bandgap energy (quantum threshold) corresponding to the visible are able to better utilize the higher quality near-UV and visible light photons than solar photovoltaic cells with bandgap energy (quantum threshold) corresponding to the near-IR even though ΔF has been normalized to the bandgap [61] and even though the narrower bandgap cells are able to absorb many more photons (refer to Figure 3.3). Optimal utilization of high-quality near-UV and visible photons becomes an important device design issue in the much higher efficiency multijunction solar photovoltaic cells that will be discussed later.

3.12 RADIATIVE RECOMBINATION COEFFICIENT AND LIFETIME

The radiative recombination coefficient B, which is used when computing radiative lifetime, is partly a function of the surrounding media [63] and may be determined by employing detailed balance [64], the same principle previously used to calculate limiting V_{oc}. Similar to the discussion in Section 3.9, this detailed balance approach may now be used to find the radiative recombination coefficient for solar photovoltaic cells with (*Case A*) a backside mirror, (*Case B*) an air-exposed backside, which is also known as a bifacial configuration, and (*Case C*) a parasitically absorbing backside substrate.

(*Case A*) The radiative recombination coefficient for a solar photovoltaic cell with a perfect backside mirror is given by:

$$B = 2\pi n_i^{-2} h^{-3} c^{-2} \int_{E_g}^{\infty} E^2 \alpha(E) \exp(-E / k_B T) dE, \qquad (3.12)$$

where n_i is intrinsic carrier concentration given by $n_i = (N_C N_V)^{1/2}$ $\exp(-E_g / 2k_B T)$ with $N_C(N_V)$ effective density of conduction (valence) band states, E is photon energy, and $\alpha(E)$ is absorption coefficient.

(*Case B*) The radiative recombination coefficient for a solar photovoltaic cell with an air-exposed backside is given by:

$$B = 2\pi(2) n_i^{-2} h^{-3} c^{-2} \int_{E_g}^{\infty} E^2 \alpha(E) \exp(-E / k_B T) dE. \qquad (3.13)$$

(*Case C*) The radiative recombination coefficient for a solar photovoltaic cell with a parasitically absorbing backside substrate is given by:

$$B = 2\pi(1+n_r^2)n_i^{-2}h^{-3}c^{-2}\int_{E_g}^{\infty} E^2\alpha(E)\exp(-E/k_BT)dE. \quad (3.14)$$

Of these three cases, the mirror cells have the smallest radiative recombination coefficient, which then leads to the longest radiative lifetime. For example, in a solar photovoltaic cell's *p*-type base layer, the radiative lifetime τ_{rad} of the minority carrier electrons is:

$$\tau_{rad} = [B\cdot(p_0+n_0+n_{ph})]^{-1}, \quad (3.15)$$

where p_0 is the equilibrium hole concentration, $n_0 = n_i^2/p_0$, and n_{ph} is the photogenerated electron concentration. If the acceptor dopants are assumed to be fully ionized at room temperature, then $p_0 \approx N_A$, where N_A represents the familiar nomenclature for acceptor doping concentration [61]. Besides radiative lifetime, we consider nonradiative lifetime next.

3.13 AUGER AND SRH LIFETIME

Auger recombination and SRH recombination are nonradiative processes that diminish solar photovoltaic cell efficiency below the ideal radiative limit. For example, in a solar photovoltaic cell's *p*-type base layer, the Auger lifetime τ_{Aug} of the minority carrier electrons is:

$$\tau_{Aug} = [C_p\cdot(p_0^2+2p_0n_{ph}+n_{ph}^2)+C_n\cdot(n_0^2+2n_0n_{ph}+n_{ph}^2)]^{-1}, \quad (3.16)$$

where C_p and C_n are the Auger recombination coefficients.

Under low injection ($n_{ph} < p_0$) in a solar photovoltaic cell's *p*-type base layer, the SRH lifetime τ_{SRH} of the minority carrier electrons is:

$$\tau_{SRH} = (\sigma_n v_{th} N_T)^{-1}, \quad (3.17)$$

where σ_n is the electron trap capture cross section, v_{th} is thermal velocity, and N_T is trap concentration. The thermal velocity is:

$$v_{th} = (3k_BT/m^*)^{1/2}, \quad (3.18)$$

where m^* is the effective mass that is typically just defaulted to the free electron mass as an expedient approximation [65]. SRH recombination depends on both the trap concentration and capture cross section.

3.14 MINORITY CARRIER DIFFUSION LENGTH

Now that we have considered recombination, one of the typical constraints in real p–n junction solar photovoltaic cells (which are known as minority carrier devices) is to design the emitter and base layers to have thickness values that are less than the minority electron and hole diffusion lengths $L_n(L_p)$ given by

$$L_n = (\tau_n D_n)^{1/2}$$
$$L_p = (\tau_p D_p)^{1/2},$$

(3.19)

where $\tau_n(\tau_p)$ is the minority electron (hole) lifetime and $D_n(D_p)$ is the minority electron (hole) diffusion coefficient, which, for a nondegenerate semiconductor, may be found from the Einstein diffusivity–mobility relationship:

$$D_n = \mu_n k_B T e^{-1}$$
$$D_p = \mu_p k_B T e^{-1},$$

(3.20)

where $\mu_n(\mu_p)$ is the minority electron (hole) drift mobility. If the minority carrier diffusion length is too small, then there is increased propensity for carriers to recombine before collection at their respective contact electrodes. Semiconductors that have large density of electronic defects and/or small charge carrier drift mobility typically have small minority carrier diffusion lengths. The Einstein diffusivity–mobility relationship involves the default assumption that the relaxed carriers are "thermally equilibrated" with the lattice and therefore it is not relevant for describing hot-carrier transport in which $T_{\text{hot carrier}} > T_{\text{lattice}}$.

3.15 MULTIPLE JUNCTIONS

Multijunction solar photovoltaic cells, a device architecture in which the widest bandgap subcell is on top and subcells with incrementally smaller bandgaps are underneath, offer a proven path for high efficiency by using the incident sunlight more effectively. An ideal (or nearly ideal) single junction solar photovoltaic cell tuned to respond to the AM1.5G spectrum would have a bandgap, as mentioned earlier and shown in Figure 3.2, of ~0.91 eV to 1.65 eV. For example, 1.42 eV direct bandgap GaAs is essentially an ideal choice for a single junction cell. However, even with this seemingly ideal choice, all the IR photons with energy less than ~1.42 eV are wasted (not absorbed), while the near-UV, visible, and near-IR photons with energy greater than 1.42 eV generate carriers that lose their excess energy as explained in Section 3.7. Therefore, more than one subcell may be configured in a stack to more selectively respond to the incident spectrum and ultimately attain superior power conversion efficiency when compared to a single junction cell alone. In other words, total carrier relaxation loss can be reduced, and with enough subcells, a broader portion of the total solar spectrum may be accessed as well.

An important point is that multijunction solar photovoltaic cells may be fabricated in a monolithic, series-connected, two-terminal configuration in which individual subcells are connected electrically, optically, and mechanically by integral tunnel (Esaki) diodes. Note that in a series-connected configuration, the nominal goal is to match the photogenerated current density, and more precisely, the maximum power point current density, in each subcell. For series-connected subcells, the subcell voltage is additive, while the current density is limited by the subcell producing the least current density. Naturally, then, as the terrestrial solar spectrum varies throughout the day, subcells become current mismatched. Nonetheless, a well-designed series-connected multijunction cell still easily outperforms a single junction cell even when factoring in subcell current mismatch.

Semiconductor layers are typically fabricated by an epitaxial growth process, for example, in either a metal organic vapor phase epitaxy (MOVPE) system or a molecular beam epitaxy (MBE) system. A detailed view of a 1.88 eV/1.42 eV 2J cell (e.g., 1.88 eV AlGaAs/1.42 GaAs) is shown in Figure 3.10. An equilibrium (dark-state) band diagram view of this 2J cell including a tunnel junction is shown in Figure 3.11. This 2J bandgap combination has a detailed balance-limiting efficiency of 39.5% (AM1.5G 1× spectrum, T_{cell} = 298 K). Note the limiting efficiency calculations here assume each subcell has a perfect backside selective reflector such as an epitaxial distributed Bragg reflector (DBR). Substantially, more efficient 6J cells will be discussed in Chapter 4.

Compound (class III–V) semiconductor multijunction solar photovoltaic cells are expensive. So far, relatively wide area (~26.6 cm²) 3J

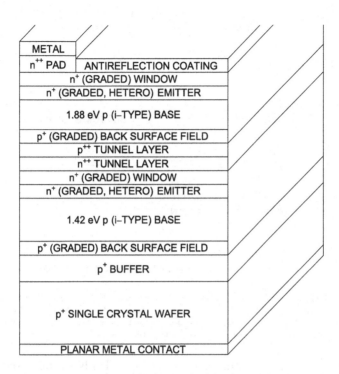

Fig. 3.10. Schematic of a 1.88 eV/1.42 eV double junction (2J) solar photovoltaic cell. Note that this figure is intended to represent a generic example of a 2J cell, and serves to illustrate the complex stack of epitaxial layers typical of compound semiconductor multijunction cells. There is flexibility in multijunction device design, such as specific composition of base, emitter, window, BSF, and TJ layers as well as doping concentration and thickness of individual layers. In addition, there is flexibility in choice of contact electrode metals and antireflection coating layer stacks. Some of the possible design variations are indicated by the content in parentheses.

cells (e.g., ~1.88 eV $Ga_{0.5}In_{0.5}P$/1.41 eV $Ga_{0.99}In_{0.01}As$/0.66 eV Ge) are routinely used in space to power satellites due to their superior beginning and end-of-life (EOL) power conversion efficiency compared to single junction Si or GaAs. Yet, for application on Earth, these expensive compound semiconductor multijunction cells are typically fabricated in sizes that are ≤ 1 cm^2 area and then incorporated into concentrated photovoltaic (CPV) modules where either refractive optics (e.g., Fresnel lenses) or reflective optics (e.g., parabolic mirrors) are used to concentrate direct beam solar radiation onto the tiny devices. The rationale here is that the optics and Sun tracking gear required to keep the small-area cells pointed directly toward the Sun throughout the day are cheaper than using wide-area multijunction solar photovoltaic cells in a flat-plate one Sun module.

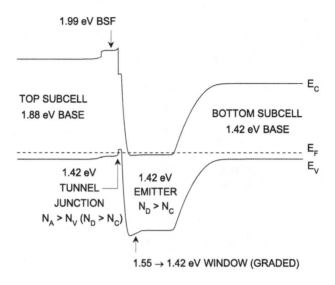

Fig. 3.11. Equilibrium (dark-state) band diagram of a 1.88 eV/1.42 eV double junction (2J) solar photovoltaic cell, showing the central tunnel junction region, in particular, generated by using ADEPT/F v2.1b [66] and the properties of $Al_xGa_{1-x}As$ [67]. Key: N_D (N_A) is donor (acceptor) doping concentration, N_C (N_V) is effective density of conduction (valence) band states, and BSF is back surface field.

Concentrating the sunlight by a factor C leads to increased power conversion efficiency where photocurrent increases as:

$$J_{ph,\,concentration} = C \cdot J_{ph,\,one\,Sun}, \tag{3.21}$$

and the open circuit voltage increases as:

$$V_{oc,\,concentration} = V_{oc,\,one\,Sun} + k_B T e^{-1} \ln C. \tag{3.22}$$

The 1.88 eV/1.42 eV 2J cell under 500× direct beam sunlight concentration (AM1.5D, T_{cell} = 298 K) has a detailed balance-limiting efficiency of 43.6% (compared to 39.5% under the AM1.5G 1× spectrum). Note that the AM1.5D spectrum has a one Sun irradiance of 0.09 W cm⁻²; therefore, the incident AM1.5D irradiance at 500× concentration (C = 500) is 45 W cm⁻². As a side note, and a point of trivia, the maximum sunlight concentration here on Earth is given by:

$$C_{max} = \frac{\pi}{\Omega_{Sun}}, \tag{3.23}$$

where Ω_{Sun} is the solid angle subtended by the Sun (~6.8 × 10⁻⁵ sr). Solving Equation 3.23 gives C_{max} ~ 46200. So far, practical sunlight concentration values for CPV application seem to occur when $C \leq 1200$. Beyond $\sim 1200\times$ sunlight concentration, the differential gain in power conversion efficiency really slows down.

3.16 HOT-CARRIER CELLS

As mentioned earlier in Section 3.2, we have not placed any special emphasis on the so-called "3rd Generation" solar photovoltaic cells simply because they have been obviated so far by the eminently proven and pragmatic multijunction cells. In this section, however, we will briefly examine the concept of hot-carrier cells, proposed in 1982 by Ross and Nozik [68], to elucidate the challenges of hot-carrier extraction. In a hot-carrier cell, the goal is to rapidly extract the photogenerated carriers at steady state when they are still described, more or less, as a thermal population with $T_{hot\ carrier} > T_{lattice}$. In this regime, as noted earlier, the standard Einstein diffusivity–mobility relationship does not apply. Therefore, as best possible, it appears that the cell should be designed such that the hot carriers transport quasi-ballistically to energy selective contacts (ESC). The electron and hole ESC – such as resonant tunnel diode structures – are designed to have a narrow energy level (i.e., nearly monochromatic) to facilitate isentropic hot-carrier extraction as shown generically in Figure 3.12.

One immediate challenge of attaining successful hot-carrier cell operation is embodied in the naturally wide distribution of the photo-generated hot carriers, which results because of the polychromatic nature of sunlight. As an example, the photogenerated hot-carrier distribution in GaAs is shown in Figure 3.13. In other words, the hot carriers are not photogenerated in a narrow energy range to begin with as would happen if they were photogenerated from monochromatic laser light. Therefore, the hot carriers must somehow be redistributed into a narrow energy range to set up successful extraction through the ESC. Recall, as mentioned in Section 3.7, the hot-carrier relaxation is rapid. As a quantitative example, hot electron and hot hole relaxation in GaAs is shown in Figure 3.14.

It has been proposed that instead of a bulk semiconductor, a superlattice structure could be used to help suppress carrier–optical phonon interaction [70]. For example, hot carriers could be photogenerated in

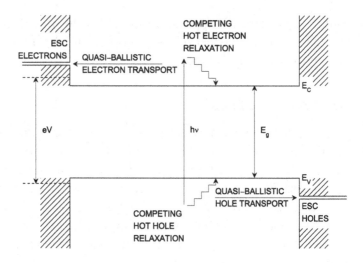

Fig. 3.12. Schematic of a hypothetical hot-carrier solar photovoltaic cell. The photogenerated hot carriers should be extracted approximately isentropically through narrow ESC, which might consist of resonant tunneling diode structures. Rapid (quasi-ballistic) transport of hot electrons and hot holes to their respective ESC would ideally dominate any competing tendency for carriers to succumb to relaxation all the way to/near the fundamental band edges. (After Ref. [69].)

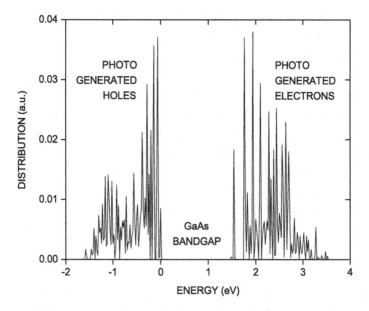

Fig. 3.13. Hot-carrier photogeneration histogram for a GaAs solar photovoltaic cell, AM1.5G spectrum. This plot depicts the photogenerated hot carriers prior to any subsequent relaxation. (After Ref. [69].)

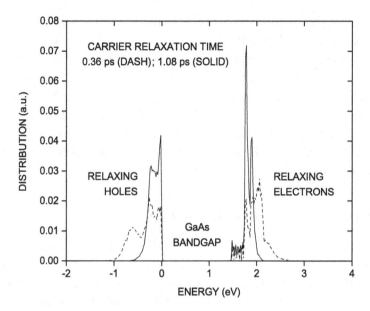

Fig. 3.14. Hot-carrier relaxation histogram for a GaAs solar photovoltaic cell, AM1.5G spectrum. Notice how rapidly (picosecond timescale) the carrier distribution changes compared to the initial photogenerated hot-carrier distribution shown in Figure 3.13. (After Ref. [69].)

a miniband of the superlattice where this miniband would be designed narrow enough in energy ($E_{\text{miniband width}} < \hbar\omega_{\text{optical phonon}}$) that emission or absorption of optical phonons would be suppressed while the gap energy between superlattice minibands would be greater than the optical phonon energy ($E_{\text{gap, miniband to miniband}} > \hbar\omega_{\text{optical phonon}}$) to prevent out-scattering of electrons [70]. Then, the desire is that the electrons in the miniband might quasi-ballistically transport to an electron ESC (and supposedly the same process would occur separately for holes in a hole miniband).

However, regardless of whether a bulk semiconductor or nanostructured semiconductor such as a superlattice is used, it appears that a simple and elegant 2J cell is a compelling alternative to the complex hot-carrier cell approach. In this sense, the energy gap that defines the separation in energy levels between the electron ESC and hole ESC in the hot-carrier cell would default to the top subcell bandgap of the 2J cell, whereas the narrower bandgap energy of the hot-carrier absorber layer itself would be equivalent to the bandgap energy of the bottom subcell of the 2J cell [69] with the caveat that the top and bottom subcell bandgaps are chosen so as to optimally "match" the particular solar spectrum that the

2J cell is designed to respond to. A near-ideal approach, for example, could be comprised of bandgaps of 1.56 eV and 0.92 eV for a 2J cell that responds well to the AM1.5D spectrum. This could, in theory, be realized with lattice-matched 1.56 eV AlGaAs and 0.92 eV GaInNAsSb, with a detailed balance-limiting efficiency of 52.9% under the AM1.5D spectrum (T_{cell} = 298 K; C = 500×). The other advantage of the series-connected 2J cell is that the current density is halved in comparison to a standalone 0.92 eV 1J cell, and this means there will be less parasitic I^2R loss.

3.17 DEVICE ENGINEERING DETAILS

Although seemingly mundane, besides the usual care needed in so far as preparing high-quality and low defect density semiconductor layers with long minority carrier lifetime, a large measure of the cell-level efficiency evolution of single and multiple junction solar photovoltaic cells is more or less driven by how well the following five losses are controlled: (i) I^2R loss, (ii) antireflection coating loss, (iii) top grid contact electrode shadow loss, (iv) front and rear surface recombination loss, and (v) parasitic backside absorption loss. Series resistance occurs in semiconductor (including transparent conductive oxide) layers, at the semiconductor–metal contact electrode interface, and in the metal contact electrodes themselves. Series resistance also occurs in the tunnel diodes that are part of multijunction cells. Good engineering practices can help to mitigate series resistance and thus I^2R loss. Typical single layer antireflection coatings and even the more advanced two-layer ARC (e.g., MgF_2 / ZnS) designed via the usual quarter-wavelength interference technique are not perfectly antireflective over the broad-band solar radiation that they must respond to (i.e., photons with energy from the near-UV to the solar cell bandgap energy), but careful design can mitigate the losses. Top grid contact electrode shadow loss can be minimized through careful attention to design and fabrication processes, or even eliminated entirely in the case of the all-back contact design configuration in which both n- and p-type contacts are located (usually in an interdigitated format) on the backside of the solar photovoltaic cell (e.g., SunPower Maxeon®). Surface recombination loss has been markedly reduced in the highest quality solar photovoltaic cells through the use of wide-bandgap passivating layers such as a–Si:H (e.g.,

Panasonic HIT®), SiO_2, $SiN_x:H$, or Al_2O_3 on c–Si cells and AlGaAs, GaInP, or AlInP on GaAs cells. Finally, as discussed in Section 3.9, a well-designed mirror can mitigate parasitic backside absorption loss that degrades the open circuit voltage V_{oc} below that for a solar photovoltaic cell that has a perfect mirror. Bifacial cell architecture (e.g., Panasonic HIT® Double) provides another excellent solar photovoltaic cell design option.

REFERENCES

[1] A.E. Becquerel, Recherches sur les effets de la radiation chimique de la lumière solaire, au moyen des courants électriques, Comptes Rendus de l'Académie des Sciences 9 (1839) 145–149.

[2] A.E. Becquerel, Mémoire sur les effets électriques produits sous l'influence des rayons solaires, Comptes Rendus de l'Académie des Sciences 9 (1839) 561–567.

[3] W. Smith, Effect of light on selenium during the passage of an electric current, Nature 7 (1873) 303.

[4] F. Braun, Ueber die Stromleitung durch Schwefelmetalle, Annalen der Physik und Chemie 229 (1875) 556–563.

[5] J.W. Gibbs, On the equilibrium of heterogeneous substances, Transactions of the Connecticut Academy of Arts and Sciences 3 (1875–1876) 108–248.

[6] E.H. Hall, On a new action of the magnet on electric currents, American Journal of Mathematics 2 (1879) 287–292.

[7] H. Hertz, Ueber einen einfluss des ultravioletten lichtes auf die electrische entladung, Annalen der Physik und Chemie 267 (1887) 983–1000.

[8] M. Planck, Ueber das gesetz der energieverteilung im normalspectrum, Annalen der Physik 309 (1901) 553–563.

[9] A. Einstein, Über einen die erzeugung und verwandlung des lichtes betreffenden heuristischen gesichtspunkt, Annalen der Physik 322 (1905) 132–148.

[10] J. Czochralski, Ein neues verfahren zur messung der kristallisationsgeschwindigkeit der metalle, Zeitschrift für Physikalische Chemie 92 (1918) 219–221.

[11] E. Wigner, F. Seitz, On the constitution of metallic sodium, Physical Review 43 (1933) 804–810.

[12] M. Riordan, L. Hoddeson, Crystal Fire, W.W. Norton, New York, 1997.

[13] R.S. Ohl, Light-Sensitive Electric Device, US Patent 2,402,662, Issued: June 25, 1946, Filed: May 27, 1941.

[14] J. Bardeen, Surface states and rectification at a metal semi-conductor contact, Physical Review 71 (1947) 717–727.

[15] W. Shockley, The theory of p-n junctions in semiconductors and p-n junction transistors, Bell System Technical Journal 28 (1949) 435–489.

[16] R.N. Hall, Germanium rectifier characteristics, Physical Review 83 (1951) 228.

[17] R.N. Hall, Electron–hole recombination in germanium, Physical Review 87 (1952) 387.

[18] W. Shockley, W.T. Read Jr., Statistics of the recombinations of holes and electrons, Physical Review 87 (1952) 835.

[19] D. Trivich, P.A. Flinn, Maximum efficiency of solar energy conversion by quantum process-es, in: F. Daniels, J.A. Duffie (Eds.), Solar Energy Research, University of Wisconsin Press, Madison, 1955, pp. 143–147.

[20] P. Rappaport, The electron-voltaic effect in *p-n* junctions induced by beta-particle bombard-ment, Physical Review 93 (1954) 246–247.

[21] D.M. Chapin, C.S. Fuller, G.L. Pearson, A new silicon *p-n* junction photocell for converting solar radiation into electrical power, Journal of Applied Physics 25 (1954) 676–677.

[22] D.C. Reynolds, G. Leies, L.L. Antes, R.E. Marburger, Photovoltaic effect in cadmium sulfide, Physical Review 96 (1954) 533–534.

[23] W. van Roosbroeck, W. Shockley, Photon-radiative recombination of electrons and holes in germanium, Physical Review 94 (1954) 1558–1560.

[24] D.A. Jenny, J.J. Loferski, P. Rappaport, Photovoltaic effect in GaAs *p-n* junctions and solar energy conversion, Physical Review 101 (1956) 1208.

[25] E.D. Jackson, Areas for improvement of the semiconductor solar energy converter, Proc. Conference on the Use of Solar Energy – The Scientific Basis 5 (1955) 122–126.

[26] E.D. Jackson, Solar energy converter, US Patent 2,949,498, Issued: August 16, 1960, Filed: October 31, 1955.

[27] C.-T. Sah, R.N. Noyce, W. Shockley, Carrier generation and recombination in *p-n* junctions and *p-n* junction characteristics, Proceedings of the IRE 45 (1957) 1228–1243.

[28] L. Esaki, New phenomenon in narrow germanium *p-n* junctions, Physical Review 109 (1958) 603–604.

[29] Naval Research Laboratory. Available from: http://www.nrl.navy.mil/.

[30] D. Kearns, M. Calvin, Photovoltaic effect and photoconductivity in laminated organic systems, The Journal of Chemical Physics 29 (1958) 950–951.

[31] P. Rappaport, The photovoltaic effect and its utilization, Solar Energy 3 (1959) 8–18.

[32] R.L. Anderson, Germanium–gallium arsenide heterojunctions, IBM Journal of Research and Development 4 (1960) 283–287.

[33] H.J. Queisser, Detailed balance limit for solar cell efficiency, Materials Science and Engineer-ing B 159–160 (2009) 322–328.

[34] W. Shockley, H.J. Queisser, Detailed balance limit of efficiency of *p-n* junction solar cells, Journal of Applied Physics 32 (1961) 510–519.

[35] Z.I. Alferov, R.F. Kazarinov, Semiconductor laser with electric pumping, USSR Patent 950840, Priority: March 30, 1963.

[36] H. Kroemer, A proposed class of heterojunction injection lasers, Proceedings of the IEEE 51 (1963) 1782–1783.

[37] H. Gerischer, M.E. Michel-Beyerle, F. Rebentrost, H. Tributsch, Sensitization of charge injec-tion into semiconductors with large band gap, Electrochimica Acta 13 (1968) 1509–1515.

[38] Z.I. Alferov, V.M. Andreev, M.B. Kagan, I.I. Protasov, V.G. Trofim, Solar-energy converters based on *p-n* $Al_xGa_{1-x}As$-GaAs heterojunctions, Fizika i Tekhnika Poluprovodnikov 4 (1970) 2378.

[39] K.W. Böer, E. Riehl, The life of the solar pioneer Karl Wolfgang Böer, iUniverse, Blooming-ton, 2010.

[40] S. Wagner, J.L. Shay, P. Migliorato, H.M. Kasper, $CuInSe_2$/CdS heterojunction photovoltaic detectors, Applied Physics Letters 25 (1974) 434–435.

[41] C.E. Backus, Solar-energy conversion at high solar intensities, Journal of Vacuum Science and Technology 12 (1975) 1032–1041.

[42] D.E. Carlson, C.R. Wronski, Amorphous silicon solar cell, Applied Physics Letters 28 (1976) 671–673.

[43] S.M. Bedair, M.F. Lamorte, J.R. Hauser, A two-junction cascade solar cell structure, Applied Physics Letters 34 (1979) 38–39.

[44] C.H. Henry, Limiting efficiencies of ideal single and multiple energy gap terrestrial solar cells, Journal of Applied Physics 51 (1980) 4494–4500.

[45] M. Taguchi, A. Yano, S. Tohoda, K. Matsuyama, Y. Nakamura, T. Nishiwaki, et al. 24.7% record efficiency HIT solar cell on thin silicon wafer, IEEE Journal of Photovoltaics 4 (2014) 96–99.

[46] M.A. Green, K. Emery, Y. Hishikawa, W. Warta, E.D. Dunlop, Solar cell efficiency tables (version 43), Progress in Photovoltaics: Research and Applications 22 (2014) 1–9.

[47] M.A. Green, Third Generation Photovoltaics: Ultra-High Conversion Efficiency at Low Cost, Springer, Berlin, 2003.

[48] A.P. Kirk, An Analysis of Multijunction, Quantum Coherent, and Hot Carrier Solar Photovoltaic Cells, PhD Dissertation, University of Texas at Dallas, 2012.

[49] W. Shockley, Electrons and Holes in Semiconductors, D. Van Nostrand Company, New York, 1950.

[50] S.M. Sze, Physics of Semiconductor Devices, 2nd ed., John Wiley & Sons, New York, 1981.

[51] D.K. Ferry, Semiconductors: Bonds and Bands, IOP, Bristol, 2013.

[52] C. Kittel, Elementary Statistical Physics, John Wiley & Sons, New York, 1958.

[53] C. Kittel, H. Kroemer, Thermal Physics, 2nd ed., W.H. Freeman, San Francisco, 1980.

[54] P.T. Landsberg, G. Tonge, Thermodynamic energy conversion efficiencies, Journal of Applied Physics 51 (1980) R1–R20.

[55] S. Adachi, Optical Constants of Crystalline and Amorphous Semiconductors: Numerical Data and Graphical Information, Kluwer Academic Publishers, Boston, 1999.

[56] D.K. Ferry, Semiconductor Transport, Taylor & Francis, London, 2000.

[57] J.R. Chelikowsky, M.L. Cohen, Nonlocal pseudopotential calculations for the electronic structure of eleven diamond and zinc-blende semiconductors, Physical Review B 14 (1976) 556–582.

[58] A.P. Kirk, Advancing solar cells to the limit with energy cascading, in: Proc. IEEE 39th Photovoltaic Specialists Conference, 2013, pp. 0782–0787.

[59] P.A. Basore, D.A. Clugston, PC1D v5. 9, University of New South Wales, Sydney, 2003.

[60] A. Martí, J.L. Balenzategui, R.F. Reyna, Photon recycling and Shockley's diode equation, Journal of Applied Physics 82 (1997) 4067–4075.

[61] A.P. Kirk, W.P. Kirk, First principle analyses of direct bandgap solar cells with absorbing substrates versus mirrors, Journal of Applied Physics 114 (2013) 174507.

[62] W.P. Dumke, Spontaneous radiative recombination in semiconductors, Physical Review 105 (1957) 139–144.

[63] E. Yablonovitch, T.J. Gmitter, R. Bhat, Inhibited and enhanced spontaneous emission from optically thin AlGaAs/GaAs double heterostructures, Physical Review Letters 61 (1988) 2546–2549.

[64] J. Nelson, The Physics of Solar Cells, Imperial College Press, London, 2003.

[65] R.F. Pierret, Advanced Semiconductor Fundamentals Volume VI, 2nd ed., Prentice Hall, New Jersey, 2003.

[66] J. Gray, M. McLennan, ADEPT/F version 2.1b, http://nanohub.org/resources/adept, 2008, doi: 10.4231/D31Z41S1T.

[67] M. Levinshtein, S. Rumyantsev, M. Shur, Handbook Series on Semiconductor Parameters Volume 2: Ternary and Quaternary III–V Compounds, World Scientific, Singapore, 1999.

[68] R.T. Ross, A.J. Nozik, Efficiency of hot carrier solar energy converters, Journal of Applied Physics 53 (1982) 3813–3818.

[69] A.P. Kirk, M.V. Fischetti, Fundamental limitations of hot-carrier solar cells, Physical Review B 86 (2012) 165206-1–165206-12.

[70] H. Sakaki, Quantum wire superlattices and coupled quantum box arrays: a novel method to suppress optical phonon scattering in semiconductors, Japanese Journal of Applied Physics 28 (1989) L314–L316.

CHAPTER 4

Energy Cascading

4.1 INTRODUCTION

This chapter commences by invoking high-efficiency combined-cycle thermal power plants (natural gas + steam turbines) as a benchmark for comparing the power conversion efficiency of advanced 6-junction (6J) concentrated photovoltaic (CPV) modules. The volumetric power densities of solar photovoltaic cells operating without sunlight concentration as well as under sunlight concentration are compared next. Then, sunlight concentration with optics is discussed along with the need for Sun tracking. After this, the limiting efficiency of optimal 6J CPV cells is examined, followed by a look at cell and module losses. The efficiency trend of the 6J CPV cells as a function of sunlight concentration is then presented. A comparison of 6J and 9-junction (9J) CPV cells is given. Finally, this chapter concludes by providing motivation for advancing the technology limits of CPV cells and modules to achieve substantial improvements in field-deployed photovoltaic power conversion efficiency.

4.2 COMBINED-CYCLE THERMAL POWER PLANTS

Combined-cycle thermal power plants utilize sufficiently high-quality heat from the exhaust of gas turbines as a means to generate steam to power steam turbines. These thermal power plants exceed 60% power conversion efficiency as of the year 2014 where 50 or 60 Hz AC electricity is output to the electricity grid. A typical combined-cycle thermal power plant is at a scale of 400+ MW. Combined-cycle thermal systems achieve such high efficiency through an energy cascading process that, in this context, means heat from the gas turbine exhaust that normally would be wasted immediately to the surrounding environment is instead used, in part, to produce steam (via a heat recovery steam generator) that drives the steam turbine.

Solar Photovoltaic Cells: Photons to Electricity. http://dx.doi.org/10.1016/B978-0-12-802329-7.00004-3

The result of this energy cascading process is manifested as superior combined-cycle system efficiency compared to a standalone gas or steam turbine-based power plant. For example, a well-designed simple-cycle gas turbine power plant might have an efficiency of 40%; therefore, a combined-cycle gas + steam turbine system with better than 60% efficiency offers a substantial improvement, thereby underscoring the value of employing an energy cascading design approach such as that embodied by combined-cycle thermal power plants. Note that gas turbines operate on the Brayton cycle while steam turbines operate on the Rankine cycle.

An example of an actual combined-cycle power plant is a 415 MW single-shaft design from Siemens, installed in South Korea as the Dangjin 3 turnkey project. This plant is fueled by liquefied natural gas (LNG) and uses a SGT6–8000H air-cooled gas turbine, SST6–5000 steam turbine, and SGen6–2000H two-pole hydrogen-cooled generator. A synchronous self-shifting clutch couples the generator and steam turbine, while the heat recovery steam generator (HRSG) contains a high-pressure once-through Benson® evaporator. This combined-cycle power plant, a 60 Hz unit, has achieved ~61% efficiency [1].

The recent, and controversial, surge in the early twenty-first century in hydraulic fracturing "fracking" technology used to extract natural gas from shale has increased the supply of natural gas (also known here as shale gas), at least in some countries such as the USA. As a result of the radical upward trend in fracking-derived natural gas along with a trend away from coal-fired power plants that produce more pollution and more CO_2 than natural gas-based power plants, interest in natural gas-fired combined-cycle thermal power plants remains strong at least as of the writing of this book. One additional reason for this interest is that solar photovoltaic panels and wind turbines often generate electricity variably or not at all due to cloudy conditions, dust or snow coverage, or nighttime in the case of solar photovoltaic panels and low or no wind conditions in the case of wind turbines. Therefore, when these renewable energy technologies are not able to generate enough electricity to meet demand, then combined-cycle power plants can provide the necessary power with rapid ramp rates and the ability to be shut down when not needed. On a side note, nuclear power plants, in contrast, cannot easily meet the rapid ramp rate condition required to fill the demand void when solar and wind systems are suddenly not generating enough electricity [2]. Moreover, the disaster at TEPCO's Fukushima Daiichi nuclear power plant in Japan in 2011 has also, at least momentarily, dampened enthusiasm for nuclear power plants.

The process of energy cascading is not only relevant to thermal power plants, but it is also applicable to multijunction solar photovoltaic cells. The uppermost and widest bandgap (quantum threshold) subcells generate power from the higher energy (and therefore higher quality) sunlight photons while the bottommost and narrowest bandgap (quantum threshold) subcells generate power from the lower energy (and therefore lower quality) sunlight photons [3]. In other words, lower energy photons that are not absorbed by wider bandgap subcells are put to use in lower bandgap subcells rather than being wasted. This is similar to the heat from the natural gas turbine exhaust being used, at least in part, to generate steam for the steam turbine rather than immediately wasting the gas turbine exhaust heat without first putting it to good use.

4.3 VOLUMETRIC POWER DENSITY

Even in locations with abundant sunlight, such as Phoenix, AZ, USA, the solar irradiance is described as diffuse. Conventional crystalline Si solar photovoltaic cells that dominate the market as of 2014 are wide-area devices, typically square or pseudo-square form factor of order 100–156 mm in width and 100–200 μm in thickness. Under illumination by sunlight with irradiance of 1000 W m^{-2} (0.1 W cm^{-2}), even the most efficient Si wafer-based cells can be characterized as having a small volumetric power density. However, sunlight may be concentrated by using refractive or reflective optics as mentioned in Chapter 3. The combination of sunlight concentration and efficient multijunction solar photovoltaic cells, surprisingly, leads to a greater volumetric power density than, for example, a high-performance Honda V-6 turbocharged Formula 1® racecar internal combustion engine [4] as shown in Figure 4.1. Concentrating the sunlight with optics to increase the volumetric power density of small solar photovoltaic cells is similar to using a turbocharger to increase the volumetric power density of small internal combustion engines.

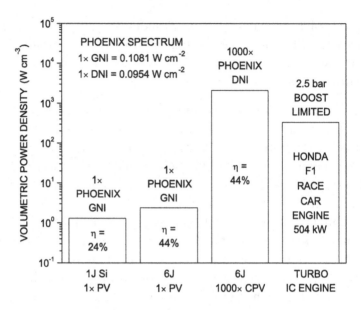

Fig. 4.1. Volumetric power density of 0.02 cm³ (1 cm² area × 200 μm thick) solar photovoltaic cells compared to a 1500 cm³ turbocharged V-6 Honda (RA168E) racecar engine. Note that the solar photovoltaic cells here are just an example; thinner and more efficient cells will yield increased volumetric power density. Key: GNI is global normal irradiance and DNI is direct normal irradiance where both of these cases assume two-axis tracking and the solar spectra are assumed to represent average conditions at Phoenix, AZ, USA. (After Ref. [3].)

4.4 SUNLIGHT CONCENTRATION WITH TRACKING

When a solar photovoltaic cell is illuminated by concentrated sunlight, the scattered (diffuse) solar radiation is wasted because only the direct beam solar radiation plus a small circumsolar component of solar radiation is accessible to the cell. When the sunlight concentration is increased, typically the acceptance angle of the collector decreases and therefore the circumsolar contribution diminishes. For high-efficiency multijunction CPV systems ($100 \leq C \leq 1200$), two-axis tracking gear keeps the cells pointed at the Sun over the seasons, from sunrise to sunset. Sunlight tracking finds an example in Nature; the approximately paraboloid-shaped flowers of *Papaver radicatum* (rooted poppy) and *Dryas integrifolia* (entire leaf mountain-avens) show Sun tracking, known technically in this context as heliotropism [5]. Although sunlight concentration is beneficial, there is a trade-off in terms of the ground-packing factor when comparing tracking CPV array fields with conventional fixed-tilt PV array fields (e.g., Si modules that are mounted with a fixed tilt angle) [6]. This trade-off in ground-packing factor is depicted in Figure 4.2.

Using the SMARTS solar spectrum modeling program discussed in Chapter 2, realistic solar spectra (global and direct) may be modeled for a specific location considering local atmospheric and weather conditions. For example, SMARTS was used to model the average global normal irradiance (GNI) and direct normal irradiance (DNI) in Phoenix, AZ, USA. In addition to the DNI, a circumsolar radiation component was modeled by assuming a pyrheliometer opening half-angle of 1.1°, which is equivalent to the effective acceptance half-angle of 1.1° of a Fresnel–Köhler optics system operating with sunlight concentration of 1000× [7]. Note that the use of a typical pyrheliometer opening half-angle of 2.5° would have overestimated the circumsolar radiation component here [8]. The Phoenix DNI is shown in Figure 4.3. This spectrum will subsequently be used to analyze the performance of a 6J CPV cell presented next. Site-specific DNI spectra are routinely utilized to investigate the performance of multijunction solar photovoltaic cells [9].

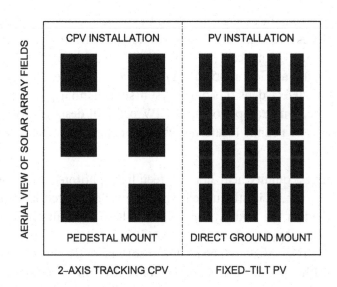

Fig. 4.2. Aerial view of a two-axis tracking CPV array field vs. a fixed-tilt PV array field. The CPV modules shown here are central pedestal mounted units. The two-axis tracking CPV modules must be spaced further apart than the fixed-tilt PV modules. The PV modules might contain Si cell technology, whereas the CPV modules might contain III–V multijunction cell technology. (After Ref. [3].)

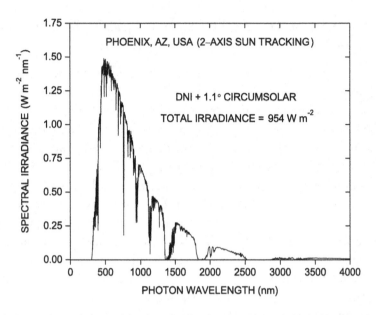

Fig. 4.3. DNI + circumsolar spectral irradiance for Phoenix, AZ, USA, as modeled with SMARTS. (After Ref. [3].)

4.5 6J CPV CELLS

6J CPV cells may be constructed with the uppermost five subcells comprised of direct bandgap III–V semiconductors that are lattice matched to the bottom Ge subcell (i.e., all subcells have nominally the same atomic spacing as Ge, or ~5.658 Å). Lattice matching allows for high-quality, low defect density single crystal subcells that provide optimal performance. An example of a 6J CPV cell [10] designed for optimal performance in a sunny location such as Phoenix, AZ, USA is shown in Figure 4.4.

The highest quality near-UV to green light photons are absorbed in the top 2.16 eV AlGaInP subcell while the yellow to red light photons are absorbed in the underlying 1.72 eV AlGaInAs subcell. The other four subcells comprised of 1.41 eV GaInAs, 1.15 eV GaInNAsSb, 0.93 eV GaInNAsSb, and 0.66 eV Ge absorb near-IR photons. Notably, under the Phoenix DNI at sunlight concentration of 1000× (or even 375×), the series-connected 6J cell described here generates 50% of its power from the top two subcells that respond to visible (and near-UV) light. The remaining four subcells that respond to the near-IR photons

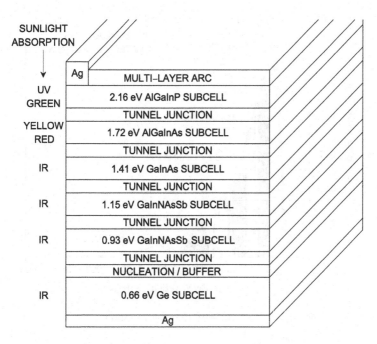

SUNLIGHT
ABSORPTION

UV GREEN	
YELLOW RED	
IR	
IR	
IR	
IR	

| Ag | MULTI-LAYER ARC |
| 2.16 eV AlGaInP SUBCELL |
| TUNNEL JUNCTION |
| 1.72 eV AlGaInAs SUBCELL |
| TUNNEL JUNCTION |
| 1.41 eV GaInAs SUBCELL |
| TUNNEL JUNCTION |
| 1.15 eV GaInNAsSb SUBCELL |
| TUNNEL JUNCTION |
| 0.93 eV GaInNAsSb SUBCELL |
| TUNNEL JUNCTION |
| NUCLEATION / BUFFER |
| 0.66 eV Ge SUBCELL |
| Ag |

Fig. 4.4. Schematic of a monolithic, lattice-matched single crystal, two-terminal, series-connected 6J CPV cell. The sunlight absorption in each subcell is also depicted. This cell may be grown with either MOVPE or MBE. Key: ARC is antireflection coating. (After Ref. [3].)

generate the remaining 50% of the power. The CPV cell that has been presented here has a top-to-bottom subcell bandgap spread of 1.5 eV (i.e., from 2.16 to 0.66 eV) – this is all that is needed to achieve high efficiency. Another way to visualize this is to look at the 6J CPV cell spectrum binning as shown in Figure 4.5. The 0.66 eV bandgap of the bottom Ge subcell has an absorption cutoff wavelength of ~1880 nm. The spectral irradiance from 1880 to 4000 nm represents less than 5% of the total for the Phoenix DNI spectrum as shown in Figure 4.6.

4.6 CELL AND MODULE LOSSES

As discussed in Chapter 3, the principle of detailed balance returns and has now been used to calculate the limiting efficiency of the 6J CPV cells presented in Section 4.5 (Figure 4.4) assuming the Phoenix DNI (Figure 4.3) for sunlight concentration of 375× and 1000× and cell temperatures of 298 K (25°C) and 373 K (100°C). The detailed balance-limiting cell efficiency is shown in Table 4.1. In reality, a variety of losses are

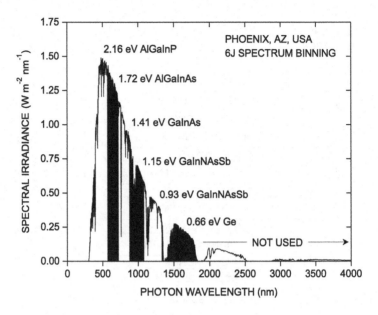

Fig. 4.5. 6J CPV subcell spectrum binning. (After Ref. [3].)

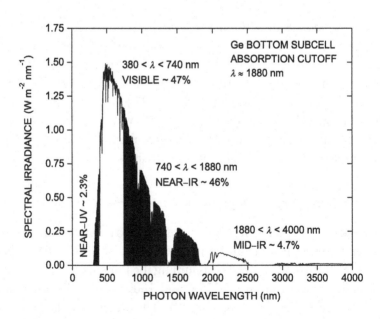

Fig. 4.6. Parsing of the Phoenix DNI spectrum into near-UV, visible, near-IR, and mid-IR spectral components.

inherent in actual solar photovoltaic cells and modules [11], and these are discussed next. First, we assume that an optimal CPV cell might achieve 86.5% of its detailed balance limit, where the losses include 4% loss due to an imperfect antireflection coating + 3% loss due to top grid contact shading + 1% parasitic I^2R loss in the emitter and window of the top subcell + 1% parasitic I^2R loss associated with the metal grid contacts and cell-to-cell interconnects + 1% parasitic I^2R loss in the tunnel diodes + 0.5% optical loss in the tunnel diodes + 3% loss due to non-radiative SRH and Auger recombination. Second, we assume that the Fresnel–Köhler optics might have an average polychromatic efficiency of 86.5% [7]. Third, we assume that the inverter has an efficiency of 98.5%. Fourth, we assume that the two-axis tracking gear has an efficiency of 98.5%. These losses result in a determination of the peak module AC efficiency. Any additional loss due to imperfect current matching among all the cells that make up the module has not been considered. In summary, the CPV peak module efficiency is given by $\eta_{module, peak} = \eta_{cell, max}$ × 0.865 × 0.865 × 0.985 × 0.985.

Note that the peak AC efficiency occurs when the 6J CPV cells are generating electricity under the exact condition that the solar spectrum (e.g., Phoenix DNI) matches the design intent (the specific spectrum such as the Phoenix DNI that ultimately fixes the design of the device such as choice of subcell bandgap). Since the solar spectrum is constantly fluctuating, the AC efficiency averaged from sunrise to sunset will be lower due to current mismatch. Meanwhile, an estimate of cell operating temperature on a hot and still day under sunlight concentration ($T_{ambient}$ = 46°C; DNI = 954 W m^{-2}) is T_{cell} ~ 100°C [12]. Peak AC module efficiency, at 25°C and 100°C cell temperature, is shown in Table 4.1. The losses outlined here are aggressive targets that are intended to help estimate, to first order, efficiency of field-deployed 6J CPV modules.

If the cells achieve less than 86.5% of the detailed balance limit, if the optics efficiency is less than 86.5%, and if the inverter and tracking efficiencies are less than 98.5%, then this will result in reduced CPV cell and module efficiencies compared to the values shown in Table 4.1. Additionally, extensive periods of cloud coverage, modest to severe dusting (soiling) of the CPV modules, and/or substantial change to the DNI due to factors such as atmospheric aerosol loading will degrade the estimated efficiency values shown here. The 6J CPV estimated peak

Table 4.1 Estimated 6J CPV Cell and Module Efficiency (375× and 1000× DNI; Phoenix, AZ, USA)

Cell Temperature	Cell Limiting Efficiency at 375×	Peak Module AC Efficiency at 375×	Cell Limiting Efficiency at 1000×	Peak Module AC Efficiency at 1000×
25°C	65.7%	47.7%	67.3%	48.9%
100°C	61.2%	44.4%	63.2%	45.9%

AC module efficiency values in Table 4.1 are not equal to the combined-cycle thermal power plants that have achieved actual efficiency of ~61%, yet they are roughly double the Si PV module efficiency record of 22.9% (1×; 25°C) as of July 2014 [13].

4.7 EFFICIENCY TREND

The detailed balance-limiting 6J CPV cell power conversion efficiency as a function of sunlight concentration is shown in Figure 4.7. From 1× to 375× sunlight concentration, there is a rapid gain in power conversion efficiency, whereas from 375× to 1000×, there is only a small gain in efficiency. Sunlight concentration greater than 375× predominantly results in an increase in the cell's volumetric power density, yet the gain

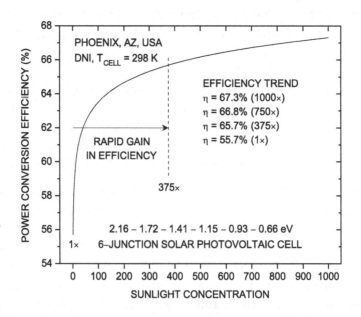

Fig. 4.7. 6J CPV cell power conversion efficiency versus sunlight concentration from 1× to 1000×. (After Ref. [3].)

in power conversion efficiency slows down and may be wasted if the parasitic I^2R power loss under larger solar concentration factors (and therefore larger photocurrents) ends up diminishing what otherwise would have been a small gain in cell power conversion efficiency.

The detailed balance-limiting figure of merit given by the change in free energy normalized to the bandgap as a function of sunlight concentration is shown in Figure 4.8 for three of the subcells of the 6J CPV cell. A rapid gain in the figure of merit $\Delta F / E_g$ is seen between 1× and 375×. As expected, the topmost 2.16 eV subcell has the best figure of merit $\Delta F / E_g$ from 1× to 1000×. Meanwhile, the bottommost 0.66 eV subcell achieves substantial increase in figure of merit $\Delta F / E_g$ by 375×, showing the benefit of sunlight concentration.

4.8 FROM SIX TO NINE SUBCELLS

Out of curiosity, we might wonder how much gain in efficiency could be achieved by adding 50% more subcells, thereby going from a 6J design to a 9J design. It turns out that the added complexity of developing a 9J CPV cell is not really worth the effort since the detailed balance-limiting

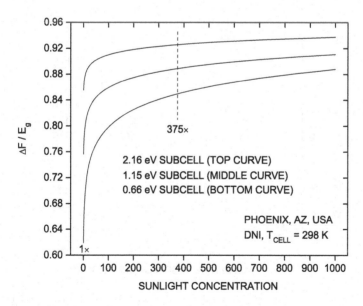

Fig. 4.8. Limiting figure of merit $\Delta F / E_g$ for 2.16 eV, 1.15 eV, and 0.66 eV current-matched subcells in a 6J CPV cell (1× to 1000 ×).

power conversion efficiency only increases from 67.3% (6J) to 69.7% (9J) under the Phoenix DNI and 1000× sunlight concentration as shown in Figure 4.9. Although a 9J CPV cell does not offer a compelling increase in efficiency versus a 6J CPV cell, a properly designed 6J CPV cell, however, is substantially more efficient than either a 1.34 eV InP 1J cell or a lattice-matched 1.92 AlGaInP/1.42 GaAs/1.04 eV GaInNAsSb 3J cell. A well-designed 6J cell offers a nearly ideal yet practical technology for achieving high power conversion efficiency. The trivial difference in limiting efficiency between the 9J and 6J cells also highlights that there is minimal benefit in designing multijunction solar photovoltaic cells with subcell bandgap energy less than 0.66 eV (λ ~ 1880 nm), as shown in Figure 4.10. It is possible to rely on proven Ge wafer substrates [14] as a platform to develop high-efficiency lattice-matched 6J cells that are nearly optimal. Therefore, there is no need to wait for alleged improvements that have been promised, but not yet delivered, from the "3rd Generation" technology to realize distinct improvements in power conversion efficiency and volumetric power density of solar photovoltaic cells and modules.

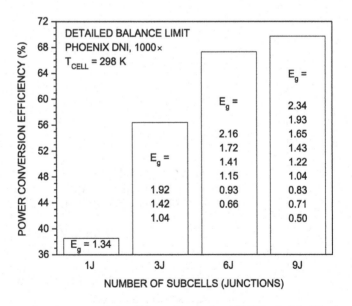

Fig. 4.9. Comparison of power conversion efficiency of 1J, 3J, 6J, and 9J CPV cells. Subcell bandgap energy is also shown. Calculation assumes all subcells have a perfect backside mirror such as a selective distributed Bragg reflector (DBR).

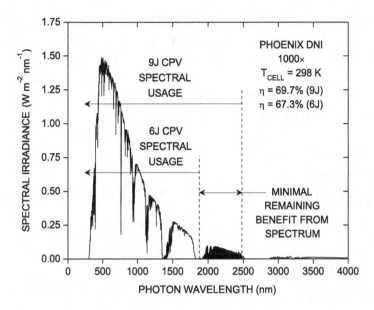

Fig. 4.10. Spectral usage of 6J versus 9J CPV cells. Accessing the solar radiation between ~1880 and 2500 nm is of little benefit.

4.9 MOTIVATION TO ADVANCE THE TECHNOLOGY LIMITS

One purpose of this chapter where we compare 6J CPV modules vs. combined-cycle thermal power plants is to provide motivation for advancing the technological limits of high-performance photovoltaics. In the gas turbine industry, the limits of the engines are being steadily advanced, allowing H-class turbines to achieve extreme firing temperature greater than 1427°C (1700 K) [15] resulting in improved thermal efficiency and, in part, ultimately leading to the ~61% efficient combined-cycle systems mentioned in this chapter. If we make an overtly simplistic assumption and invoke the Carnot limit, the idealized peak efficiency of a combined-cycle power plant with 1700 K gas turbine firing temperature and 300 K exhaust from the steam turbine is ~82%; therefore, there is still opportunity to realize >61% efficiency in actual combined-cycle power plants. As of July 2014, the record CPV module efficiency was only 36.7% (4J wafer-bonded CPV cells under 230× light concentration) indicating that CPV has a substantial gap to close yet to reach an efficiency value reasonably on par with combined-cycle thermal power plants. Similar to the state-of-the-art combined-cycle power plants, it will take comprehensive and advanced engineering to develop and certify the best performing, most reliable, and cost-competitive CPV technology that is able to reduce the need not only for coal-fired power plants but also for fracking of shale to extract natural gas for gas turbine-based power plants. As CPV scales up, though, attention should be placed on designing for recyclability and reuse of materials when the modules reach the end of their service life.

REFERENCES

[1] R. Prandi, South Korea CCPP exceeds 60% efficiency, Diesel & Gas Turbine Worldwide (March 2014) 22–23.

[2] A. Pickard, G. Meinecke, The future role of fossil power generation, Siemens AG Technical Report Order No. E50001-G220-A137-X-4A00, Erlangen, 2011, 1–20.

[3] A.P. Kirk, Advancing solar cells to the limit with energy cascading, in: Proc. IEEE 39th Photovoltaic Specialists Conference, 2013, pp. 0782–0787.

[4] Y. Otobe, O. Goto, H. Miyano, M. Kawamoto, A. Aoki, T. Ogawa, Honda Formula One turbo-charged V-6 1.5L Engine, SAE Technical Paper Series, 1989, pp. 1–8.

[5] P.G. Kevan, Sun-tracking solar furnaces in high arctic flowers: significance for pollination and insects, Science 189 (1975) 723–726.

[6] A. Luque, Will we exceed 50% efficiency in photovoltaics?, Journal of Applied Physics 110 (2011) 031301-1–031301-19.

[7] P. Benítez, J.C. Miñano, P. Zamora, R. Mohedano, A. Cvetkovic, M. Buljan, et al. High performance Fresnel-based photovoltaic concentrator, Optics Express 18 (2010) A25–A40.

[8] C.A. Gueymard, Spectral circumsolar radiation contribution to CPV, AIP Conference Proceedings 1277 (2010) 316–319.

[9] S.P. Philipps, G. Peharz, R. Hoheisel, T. Hornung, N.M. Al-Abbadi, F. Dimroth, et al. Energy harvesting efficiency of III–V triple-junction concentrator solar cells under realistic spectral conditions, Solar Energy Materials and Solar Cells 94 (2010) 869–877.

[10] R.E. Jones-Albertus, P. Misra, M.J. Sheldon, H.B. Yuen, T. Liu, D. Derkacs, et al., High efficiency multijunction solar cells, US Patent 8,697,481, Issued: April 15, 2014, Filed: December 7, 2012.

[11] R.R. King, D. Bhusari, D. Larrabee, X.-Q. Liu, E. Rehder, K. Edmondson, et al. Solar cell generations over 40% efficiency, Progress in Photovoltaics: Research and Applications 20 (2012) 801–815.

[12] F. Almonacid, P.J. Pérez-Higueras, E.F. Fernández, P. Rodrigo, Relation between the cell temperature of a HCPV module and atmospheric parameters, Solar Energy Materials and Solar Cells 105 (2012) 322–327.

[13] M.A. Green, K. Emery, Y. Hishikawa, W. Warta, E.D. Dunlop, Solar cell efficiency tables (version 44), Progress in Photovoltaics: Research and Applications 22 (2014) 701–710.

[14] G. Létay, C. Baur, A.W. Bett, Theoretical investigations of III–V multi-junction concentrator cells under realistic spectral conditions, in: Proc. 19th European Photovoltaic Solar Energy Conference, Paris, 2004.

[15] B. Haight, New air-cooled H-class, Diesel & Gas Turbine Worldwide (May 2014) 30–34.

CHAPTER 5

Resource Demands and PV Integration

5.1 INTRODUCTION

This chapter focuses on some of the resource demands (particularly semiconductor material) that come into play when solar photovoltaics (PV) are scaled to a more meaningful level than that of 2014. The scalability of Si solar photovoltaics is examined, followed by CdTe and $CuIn_xGa_{1-x}Se_2$ solar PV. Then, the scalability of Ge and GaAs-based concentrated photovoltaics (CPV) is examined. The issue of soft costs that arise during the actual deployment of PV modules is presented. Next, there is an outline of some solar energy storage technologies including electric vehicles, followed finally by a brief discussion of a flexible and evolving electric grid.

5.2 SCALABILITY OF SILICON-BASED PV

From Chapter 1, we found that solar electricity is a trivial component of overall global electricity demand as of the year 2014. If we would like to initially scale solar electricity up to a level of 1 TW (1×10^{12} W) of peak power, then it is helpful to understand how much raw material we will need, at least as a rough first order estimate. Monocrystalline and multicrystalline Si solar photovoltaic cells dominate the solar electricity marketplace in 2014. World record efficiency for a monocrystalline Si-based module is 22.9% as of June 2014 [1]. If we assume that mass production monocrystalline Si modules industry-wide will soon reach 23% average efficiency (P_{out} = 0.023 W cm^{-2} at 25°C under the AM1.5G 1× spectrum) with 100 μm thick Si wafers (ρ_{Si} = 2.33 g cm^{-3}), then ~1 × 10^9 kg (one million metric tons) of Si would be required to achieve the goal of 1 TW$_{peak}$ power. This calculation of course is ideal in that it assumes 100% yield and no kerf loss (e.g., due to the saw cutting process), no perimeter loss (e.g., due to squaring of circular cross section boules), and no etching loss (e.g., due to removal of saw damage

Solar Photovoltaic Cells: Photons to Electricity. http://dx.doi.org/10.1016/B978-0-12-802329-7.00005-5

and for creating surface texturing) – in other words, all of the raw Si is converted perfectly into Si modules (panels) that achieve 23% efficiency (which means that the cells themselves might consistently be ~25% efficient). Note that a 100 mm × 100 mm square form factor Si wafer that is 100 μm thick will have a mass of ~2.33 g. Therefore, the fabrication of 23% efficient Si modules would require a quantity of ~4.3 × 10^{11} of the 100 mm × 100 mm × 100 μm dimension Si wafers to reach the goal of 1 TW$_{peak}$. (Production Si cells may be 125 mm × 125 mm or greater in area, rather than the example here.) Note that 50 μm thick Si cells and 25% efficient Si modules would require ~4.7 × 10^8 kg of Si to reach 1 TW$_{peak}$ with the same assumptions.

At least as of 2014, it seems that the most efficient Si cells are engineered from n-type monocrystalline Si wafers. These wafers are sliced from boules grown by the Czochralski process. At first order, if we assume a diamond wire saw with core wire diameter of 100 μm [2] and Si wafers with an average thickness of 100 μm, then about half or more of the high-purity Si boule ends up as kerf i.e., "saw dust" loss. Hopefully, industry is recycling the Si kerf; otherwise, this would be a massive waste. Since the saw cutting process also damages both sides of the wafer, an additional loss is incurred upon the etch process that is used to remove the portion of the Si wafer damaged during the sawing process.

The primary point is that even with high performance and class-leading Si cell and module technology, the magnitude of ultra-high-purity Si needed to manufacture a modest value of 1 TW$_{peak}$ is already stunning. Less efficient and/or thicker Si cells coupled with manufacturing yields less than 100% increase the amount of Si, and therefore this underscores the value of developing the thinnest and most efficient cells possible that are also processed with the highest yield, most consistent binning, and least energy demand during manufacturing. Superior power conversion efficiency, alone, has a knock-on effect in terms of being able to reduce balance of systems costs and materials such as glass cover sheet, aluminum framing for the modules (frameless modules are an alternative option), wiring harnesses, junction boxes, inverters (or microinverters), and mounting hardware such as brackets, clips, and bolts. Superior power conversion efficiency as well as use of the least amount of material may also reduce the energy payback time. The best possible efficiency at reasonable cost also means that less land or rooftop area is required to generate the desired amount of power (or better yet, energy which is a more meaningful metric). In other words, simply flooding the market with mediocre efficiency (and often mediocre quality) Si modules instead of the highest efficiency Si modules represents suboptimal deployment of solar photovoltaic technology as well as a suboptimal usage of raw materials and the energy required to process them into finished solar photovoltaic cells and modules.

It should be noted that the conversion of raw SiO_2 (quartz) to solar-grade Si has its own associated energy costs and pollution, and therefore may not be as benign and green as we might initially imagine. To first produce metallurgical-grade Si, the overall carbothermic reaction $SiO_2 + 2C \rightarrow Si + 2CO$ takes place in an arc furnace at $\sim 1900°C$ (noting that intermediate reaction steps have not been shown here). Metallurgical-grade Si ($\sim 98\%$ pure) must then be converted to solar-grade Si (i.e., a more pure form of Si). First, the metallurgical-grade Si is reacted with HCl (hydrogen chloride) to form $SiHCl_3$ (trichlorosilane), which is subsequently distilled to achieve greater purity. Then, via, for example, the trichlorosilane Siemens process, $SiHCl_3$ gas in the presence of H_2 (hydrogen gas) decomposes in a reactor chamber at $\sim 1150°C$ resulting in the vapor deposition of polycrystalline Si on to electrically-heated polycrystalline Si rods that grow in diameter to yield solar-grade polycrystalline Si at 99.9999% to 99.9999999 (i.e., 6N–9N) purity. Note that HCl, H_2, SiH_2Cl_2 (dichlorosilane), and $SiCl_4$ (silicon tetrachloride) are intermediaries or byproducts of the Siemens process and may be reused, recycled, or converted, e.g., $3SiCl_4 + 2H_2 + Si \rightarrow 4SiHCl_3$ (which is known as hydrochlorination). The complex and inefficient Siemens process is estimated to require ~ 90 kWh energy per kg Si [3]. Additional energy is required to fabricate (mono or multicrystalline) Si wafers and then ultimately the actual solar photovoltaic cells and modules that require glass cover sheet and typically an aluminum frame [4]. Note that an alternative to the Siemens process is provided by fluidized bed reactor (FBR) technology that employs silane (SiH_4) gas vapor deposition onto small granules of Si that then increase in size; these larger granules of Si are removed from the reactor chamber and new starter granules of Si are introduced in a more or less continuous process. REC Silicon indicates that FBR technology uses 80–90% less energy than the conventional Siemens process [5].

5.3 SCALABILITY OF CADMIUM TELLURIDE-BASED PV

Now, we will consider CdTe thin-film cells. The world record CdTe module efficiency as of June 2014 is 17.5% [1]. If we assume 20% efficient modules (P_{out} = 0.020 W cm^{-2} at 25°C under the AM1.5G 1× spectrum) with just 1 μm thick films, the amount of CdTe (ρ_{CdTe} = 5.86 g cm^{-3}) required to achieve the goal of 1 TW$_{peak}$ is ~2.9 × 10^7 kg assuming 100% yield and no waste. The element Te appears to be the limiting material in the context of scalability of CdTe solar photovoltaic cells. The atomic mass of Te is 128 amu while the atomic mass of Cd is 112 amu, and thus Te represents 53.3% of the mass of binary zinc blende CdTe. Therefore, ~1.6 × 10^7 kg of Te would be needed to meet the 1 TW$_{peak}$ target.

According to the *USGS Mineral Commodity Summaries* report, the estimated global Te reserves is 2.4 × 10^7 kg noting that this data assumes only the Te extracted from copper (Cu); more specifically, the recoverable Te found in unrefined Cu anodes [6]. Module efficiency less than 20%, CdTe absorber layer thickness greater than 1 μm, and/or process and manufacturing yield less than 100% will add to the amount of Te needed to reach 1 TW$_{peak}$. Refinery output data for Te was withheld by the USA in 2013. The combined refinery output of Te from Canada, Japan, and Russia in 2013 was only 9.5 × 10^4 kg. However, additional countries such as China did not report refinery output data for Te in 2013 [6]. It, therefore, is difficult to forecast the true future outlook for CdTe solar photovoltaic cells. On a side note, the dominant CdTe PV company, First Solar, plans to introduce TetraSun™ crystalline Si modules based on *n*-type Si wafers and Cu (instead of Ag) contact electrodes with cell efficiency of ~21% in 2014 [7].

5.4 SCALABILITY OF CIGS-BASED PV

Another polycrystalline thin-film solar photovoltaic technology, besides CdTe, is the indium-containing compound semiconductor, $CuIn_xGa_{1-x}Se_2$ (CIGS). The bandgap energy of CIGS may be tuned from ~1.04 eV for $CuInSe_2$ (CIS) to ~1.68 eV for $CuGaSe_2$ (CGS) noting that a typical composition of CIGS for solar photovoltaic cells involves a Ga/(Ga + In) ratio of ~0.2–0.3. As a baseline approximation, we will assume that CIGS cells are instead comprised of $CuInSe_2$ and then focus our attention on

In that is thought to be the limiting raw material in the context of CIGS cells. The atomic mass of In is 115 amu, while the atomic mass of Cu is 64 amu and the atomic mass of Se is 79 amu. Indium represents ~34.1% of the mass of chalcopyrite $CuInSe_2$. If we assume 20% efficient modules (P_{out} = 0.020 W cm^{-2} at 25°C under the AM1.5G 1× spectrum) with 1 μm thick films, then the amount of $CuInSe_2$ (ρ_{CIS} = 5.77 g cm^{-3}) required to achieve the goal of 1 TW$_{peak}$ is ~2.9 × 10^7 kg assuming 100% yield and no waste. Therefore, ~9.8 × 10^6 kg of In would be needed to achieve the 1 TW$_{peak}$ goal. Similar to the prior discussion about CdTe modules, if CIGS module efficiency is less than 20%, cell thickness is greater than 1 μm, and yield is less than 100%, then considerably more In would be required. The substitution of ~20–30% Ga for In (to give a typical composition of CIGS cells) does not substantially reduce the demand for In. According to the *USGS Mineral Commodity Summaries* report, in 2013, the global refinery production of In was only 7.7 × 10^5 kg. Global reserve estimates were not available. It is worth noting that In is also in demand for applications other than CIGS modules such as LEDs, transistors, and also electrical contact layers [6].

5.5 SCALABILITY OF GERMANIUM-BASED CPV

In Chapter 4, we examined an advanced, high-efficiency 6J CPV cell technology based on Ge wafer substrates (ρ_{Ge} = 5.32 g cm^{-3}). The Ge substrates are typically around 200 μm thick, whereas the III–V epitaxial layers are about 2 μm thick per subcell for the upper five subcells, and so the Ge wafers represent the dominant semiconductor component. The 6J cell-based CPV modules that were estimated, in an optimal sense in Chapter 4, to achieve ~46% peak efficiency (Phoenix DNI) at 1000× sunlight concentration and 100°C cell temperature correlate to ~44 W power output per 1 cm^2 cell. To achieve 1 TW$_{peak}$ of solar electricity, ~2.4 × 10^6 kg of Ge would be needed, assuming perfect yield. According to the *USGS Mineral Commodity Summaries* report in February 2014, worldwide refinery production of Ge in 2014 is estimated to be only 1.55 × 10^5 kg. However, USA production values were withheld and country-by-country reserves were not available. It was reported that China has an annual Ge production capacity of ~2 × 10^5 kg. Moreover, it is thought that the USA reserves of zinc (from which Ge is often extracted) may contain 2.5 × 10^6 kg of Ge [6].

Nonetheless, the production of 1 TW_{peak} of the high-performance Ge-based 6J CPV cells and modules might be pushing the limits of Ge production, especially since Ge is also in demand for optical fibers, 2–12 μm IR optics (windows and lenses), transistors (typically using SiGe alloys), phase change materials (such as GeSbTe), polymerization catalysts, X-ray monochromators, and multijunction space solar photovoltaic cells (e.g., 3J cells based on GaInP, GaInAs, and Ge subcells). Notably, it has been reported that only 3% of Ge in zinc concentrates is recovered presently [6]. Perhaps a significant effort at improving this yield would expand the availability of Ge. If demand for Ge increases, it is likely that innovation will improve Ge extraction. Rather than just dumping the tailings from Zn mines, it would make more sense to judiciously extract as many critical elements, such as Ge and Ga, if this can be done in an economically and environmentally sensible way. In addition, if the reserves of Ge listed in the *USGS Mineral Commodity Summaries* report as "N/A" [6] turn out to be substantial, or if new deposits are found or better recycling techniques developed, then there may be enough Ge to sustain a large CPV industry as well as fulfill demand for other uses.

5.6 SCALABILITY OF GALLIUM ARSENIDE-BASED CPV

If Ge ends up as a limiting material for widely scaled CPV cells, an alternative approach could be to use GaAs wafer technology. Solar Junction, a CPV manufacturer specializing in direct bandgap dilute nitride GaInNAsSb materials, has reported the ability to epitaxially grow GaInNAsSb with a bandgap at least as small as 0.80 eV that is lattice matched with GaAs [8]. Lattice-matched 1.99 eV AlGaInP/1.50 eV AlGaAs/1.15 eV GaInNAsSb/0.80 eV GaInNAsSb 4J CPV cells [9], as shown in Figure 5.1, could be developed with detailed balance-limiting efficiency of 58.4% at 100°C cell temperature (1000× Phoenix DNI) compared to the Ge-based 6J CPV cell limiting efficiency of 63.2% at 100°C cell temperature (1000× Phoenix DNI) presented in Chapter 4. If we estimate that the peak AC efficiency of these GaAs-based 4J CPV modules is ~42% (1000× Phoenix DNI; T_{cell} = 100°C), correlating to 1 cm² area cells with a power output of ~40 W cm⁻², and fabricated on 250 μm thick GaAs wafers (ρ_{GaAs} = 5.32 g cm⁻³), then this corresponds to ~3.3 × 10⁶ kg of GaAs to achieve the goal of 1 TW_{peak} of solar electricity. Gallium is believed to be the limiting element in the context of

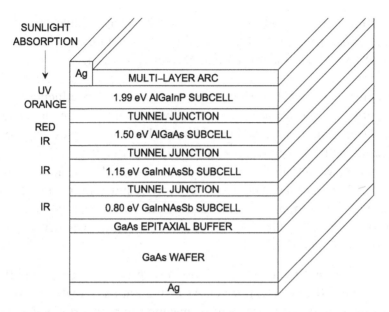

SUNLIGHT
ABSORPTION

UV
ORANGE

RED
IR

IR

IR

Ag | MULTI–LAYER ARC
1.99 eV AlGaInP SUBCELL
TUNNEL JUNCTION
1.50 eV AlGaAs SUBCELL
TUNNEL JUNCTION
1.15 eV GaInNAsSb SUBCELL
TUNNEL JUNCTION
0.80 eV GaInNAsSb SUBCELL
GaAs EPITAXIAL BUFFER
GaAs WAFER
Ag

Fig. 5.1. Schematic of a monolithic, lattice-matched single crystal, two-terminal, series-connected 4J CPV cell developed on a GaAs wafer substrate. No Ge is required and the AlGaInP and AlGaAs upper two subcells only require a small Al fraction. The sunlight absorption in each subcell is also depicted. This cell may be grown with either MOVPE or MBE. The limiting efficiency of this cell is 61.8% at T_{cell} = 25°C (58.4% at T_{cell} = 100°C) under the Phoenix DNI spectrum and 1000× sunlight concentration. Copper (Cu) contact electrodes with suitable diffusion barriers (e.g., TiN, W, etc.) could be used in place of Ag. Key: ARC is antireflection coating.

using GaAs wafers for solar photovoltaic applications. The atomic mass of Ga is 70 amu versus the atomic mass of As at 75 amu, which means that Ga is ~48.3% of the mass of binary zinc blende GaAs. Therefore, ~1.6 × 10⁶ kg of Ga would be needed to reach 1 TW$_{peak}$, assuming perfect yield and no waste throughout manufacturing.

According to the *USGS Mineral Commodity Summaries* report, global primary Ga production in 2013 was estimated at 2.8 × 10⁵ kg. It is worth noting that estimates of global Ga reserves appear to be difficult to ascertain; however, it is thought that ~1 × 10⁹ kg of Ga may exist globally in bauxite though it is unknown how much can or will be recovered [6]. Meanwhile, Ga has seen increasing demand for GaInN-based light emitting diodes (LEDs) that are displacing incandescent and fluorescent lighting, GaN-based power electronics, and recently even InGaZnO (IGZO) thin-film transistors that are displacing suboptimal amorphous Si thin-film transistor technology [6]. Ultimately, even just 1 TW$_{peak}$ of efficient GaAs wafer-based CPV puts a large burden on resource allocation, mining, and crystal growth. Preparing a huge quantity of GaAs wafers promotes a legacy of dealing with enormous quantity of toxic As and the burdens on the miners and processing personnel. Even though some of the epitaxial subcells, which are of order 1–3 μm thick, are comprised of III–As(Sb) compounds, perhaps producing a large quantity of 200 μm thick Ge wafer substrates for ultra-high-efficiency 6J CPV cells discussed in Chapter 4 would be a more ecologically sound choice compared to producing an equally large quantity of 200–250 μm thick GaAs wafer substrates especially since solar PV is supposed to be a "green" technology. This issue certainly requires more thought and consideration. Summarizing this discussion, a visual comparison of the estimated Si, Te, In, Ga, and Ge mass requirements to achieve 1 TW$_{peak}$ is shown in Figure 5.2.

It is worth noting here that it is often suggested that epitaxial lift-off could reduce the use of either GaAs or Ge wafers (growth substrates). However, just to sustain any sort of meaningful production volume and

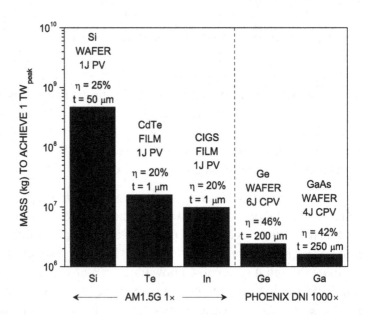

Fig. 5.2. Estimated mass requirement of Si, Te, In, Ge, and Ga to achieve 1 TW$_{peak}$ of solar electric power as a function of module efficiency (η) and cell thickness (t) noting that the primary thickness component of Ge and GaAs wafer-based multijunction CPV cells is the (200 μm thick Ge or 250 μm thick GaAs) wafer itself rather than the epitaxial subcells. It is important to consider that reduced module efficiency, increased cell thickness, and consideration of yield loss and kerf loss (e.g., for Si, Ge, and GaAs wafer-based cells) will result in the need for substantially more material.

throughput, a large quantity of wafers have to continually be available anyway, and following each epitaxial growth run and lift-off procedure (which itself is slow and may involve large quantities of etch chemicals), the wafers must then be prepared for the next epitaxial growth, which typically involves removing 10 μm or more of the wafer surface in a chemical mechanical polishing (CMP) process [10]. Besides managing this waste, another challenge is trying to maintain consistent quality and tightly binned power conversion efficiency values of the sequential batches of solar photovoltaic cells as the wafers are "refurbished" and repeatedly reused.

For In-containing III–V compound semiconductor CPV subcells lattice matched with Ge (or GaAs), we will assume, to first order, the density to be equivalent to Ge (ρ = 5.32 g cm^{-3}); for a comparative example, $Ga_{0.5}In_{0.5}P$ lattice matched with Ge has a density of 4.48 g cm^{-3} [11]. In addition, as an estimate, we will assume that any In-containing subcells are comprised of 50% In by mass. In the 6J Ge-based CPV cells discussed in Chapter 4 and earlier in this chapter, we will assume the upper five subcells that contain In are each 3 μm thick for a total of 15 μm, again as a rough estimate. To reach 1 TW_{peak} with the 44 W cm^{-2} cells would require ~9 × 10^4 kg of In. It appears to make more sense to use In for multijunction CPV solar photovoltaic cells rather than one Sun CIGS-based solar photovoltaic cells, which requires substantially more In. Note that Zn ores contain In [6]; therefore, it seems that a dedicated focus on processing Zn ores and mineral compounds (and perhaps old Zn mine tailings) may be a way to increase the amount of available In as well as Ga and Ge. All told, this analysis highlights the tremendous material demands and concerns we face as solar PV is scaled [12–14].

5.7 SOFT COSTS

As is typical for large enough volume manufacturing and the associated "economies-of-scale", the cost of (crystalline Si and polycrystalline CdTe) solar photovoltaic cells and modules has decreased notably in the earliest part of the twenty-first century. However, a large, if not dominant, component of the total expenditure involved with actually deploying PV modules is found in the so-called "soft costs" that include inspection and permitting fees, installation labor fees and installer overhead, grid interconnection fees, financing fees, module packaging and shipping fees, sales taxes and tariffs, inefficient marketing and sales processes, and operation and maintenance fees. Many more details about soft costs, and the dedicated efforts and practices underway to reduce them, may be found, for example, through the U.S. Department of Energy's Sun-Shot initiative program [15].

5.8 SOLAR ENERGY STORAGE

During nighttime, cloudy conditions, or when PV modules may be shaded or covered by dust or snow, it helps to have the capability for storing solar electricity that was not under demand in real time when the PV modules were illuminated by the Sun and generating solar electricity. Without storage of the excess solar electricity for use at nighttime, and so forth, one default may be to resort to fossil or biofuel power plants (e.g., natural gas, biogas, combined-cycle) to provide electricity when PV is not producing. Other options could be hydroelectric power plants, geothermal power plants, wind turbines, or fuel cells. Undoubtedly, the topic of solar electricity storage itself requires an entire book to extract any reasonable level of comprehension. Nonetheless, two of the possible ways to store the excess energy generated from solar photovoltaic modules are outlined next.

The excess PV-generated electricity may be used to operate an electrically powered pump or reversible water turbine that pumps water from a lower elevation lake or reservoir (or even a river) to a higher elevation lake or reservoir, thereby increasing the potential energy of the water. When demand for electricity arises, the water in the upper reservoir is allowed to flow downhill through a water turbine (e.g., a Francis or Kaplan turbine) such that the kinetic energy of the flowing water is converted to electricity via a generator (driven by the water turbine) in a process known as hydroelectric pumped storage. It should be pointed out that in many regions of the world where solar photovoltaic modules are, or will be, installed, there simply is no convenient option of using hydroelectric pumped storage to begin with. Also, just because pumped hydro storage could be used, damming rivers to create reservoirs often has a number of adverse consequences (such as displacing people from their homes, submerging historically and architecturally significant sites, or destroying habitat for fish and other wildlife and flora). Moreover, in areas with extensive and extended draught, hydroelectric pumped storage may not be advantageous to begin with.

Alternatively, the excess PV-generated energy may be used to charge rechargeable batteries that may then be discharged when electricity is needed. Of course here, as with the precious semiconductor material in the solar photovoltaic cells themselves, attention to raw material abundance [4] and allocation should be taken into consideration upon any decision to scale up the mass production of an enormous quantity of batteries.

There has been some discussion that if battery-powered electric vehicles begin to go mainstream in the twenty-first century and therefore scale in volume, then they could be used as a means to actually store excess solar electricity. While the vehicle is parked during the day, its batteries could be charged by solar photovoltaic modules (assuming once again that there is sufficient sunlight on any given day). If sufficient excess charge remains in the vehicle's batteries at night following daytime driving, then in theory, the vehicle's batteries could be partially discharged to provide electricity during the night for use at home. If there are multiple days in a row with cloudy, rainy, or snowy conditions during which PV modules are unable to provide adequate vehicle battery recharging during the day at the workplace where the vehicle may be parked during the workweek or during the day at home where the vehicle may be parked during the weekend, then it may be challenging to always rely on a strictly PV-charged electric vehicle to serve as an electricity provider at nighttime. However, it is certainly worthwhile to try and extract two benefits from an electric vehicle: a means of transportation on the one hand and a means of storing solar energy for later use on the other hand. This represents good engineering practice – obtaining two or more functions or end uses from a single design. Also, at least some of the knowledge (such as manufacturing techniques, supplier logistics, system packaging, thermal management, testing, certification, and safety-related issues) gained from researching and developing improved batteries for electric vehicles might possibly be leveraged for the improvement of solar electricity storage batteries, or vice versa. To be clear, this discussion is naïve in that (a) it assumes the majority of humans can afford and ultimately will own electric vehicles and (b) that they live in their own houses with the convenience and infrastructure of an attached garage or else have access to some other suitable place to plug in and connect their e-cars to the electrical system that services their particular residence. Moreover, and finally here, it is worth noting that both of the storage techniques discussed in this section (pumped hydro and batteries) have losses, and this should be taken into due consideration.

5.9 ELECTRIC GRID EVOLUTION

As the deployment of grid-tied solar photovoltaic modules scale up, it seems evident that the established legacy electric grid must evolve. Solar electricity output varies throughout the day as the solar spectrum changes from sunrise to sunset. Also, fluctuation in daily temperature impacts the PV module solar electricity output. Finally, intermittent cloud cover also affects PV module solar electricity output. This variable nature of solar electricity poses challenges to utility companies. If PV is grid-tied, then the real time solar electricity should be controllably conditioned and fed on to the electric grid in such a way as to reliably and safely meet customer demand. However, if at any given time the PV modules happen to be generating electricity in excess of the demand from grid-connected customers, then the solar energy could, instead, be stored for later use (e.g., during cloudy conditions or during the night). This balancing act of electricity management requires intelligent control with the proper hardware and software.

Solar electricity coupled with an efficient battery storage system that includes electricity demand buffering can reduce monthly electricity bills because electric companies charge not only a base "energy" charge but also a "demand" charge. For example, spikes in power that might occur when a business intermittently operates, say, an industrial motor-driven process at the same instance that an air-conditioning (A/C) system is cycled on during a particularly hot day can cause a substantial increase to its monthly electricity bill even though the electricity demand otherwise is well under those intermittent and perhaps even infrequent moments of demand spikes that were incurred from operating the industrial motor while cycling on the A/C. Sharp Electronics Corp. has recently (in 2014) introduced the turnkey SmartStorage system comprised of lithium (Li) ion (specifically lithium manganese oxide) storage batteries with predictive controllers, site metering, and online monitoring so that electricity generated from solar photovoltaic panels may be stored in the Li ion batteries that may then be discharged quickly in response to an imminent demand spike [16]. This results in a reduced demand charge and therefore a less costly monthly electric bill while also providing clean solar electricity to help meet demand locally on site rather than having to, for example, burn more coal in a remotely sited coal-fired power plant.

REFERENCES

[1] M.A. Green, K. Emery, Y. Hishikawa, W. Warta, E.D. Dunlop, Solar cell efficiency tables (version 44), Progress in Photovoltaics: Research and Applications 22 (2014) 701–710.

[2] H. Forstner et al., International Technology Roadmap for Photovoltaic, Fifth Edition, 2014.

[3] A. Goodrich, P. Hacke, Q. Wang, B. Sopori, R. Margolis, T.L. James, et al. A wafer-based monocrystalline silicon photovoltaics road map: utilizing known technology improvement opportunities for further reductions in manufacturing costs, Solar Energy Materials and Solar Cells 114 (2013) 110–135.

[4] M. Tao, Terawatt Solar Photovoltaics: Roadblocks and Opportunities, Springer, London, 2014.

[5] REC Silicon. Available from: http://www.recsilicon.com.

[6] USGS Mineral Commodity Summaries. Available from: http://minerals.usgs.gov/minerals/pubs/mcs/.

[7] First Solar. Available from: http://www.firstsolar.com/.

[8] V. Sabnis, H. Yuen, M. Wiemer, High-efficiency multijunction solar cells employing dilute nitrides, AIP Conference Proceedings 1477 (2012) 14–19.

[9] R.E. Jones-Albertus, P. Misra, M.J. Sheldon, H.B. Yuen, T. Liu, D. Derkacs, et al., High Efficiency Multijunction Solar Cells, US Patent 8,697,481, Issued: April 15, 2014, Filed: December 7, 2012.

[10] A.P. Kirk, An Analysis of Multijunction, Quantum Coherent, and Hot Carrier Solar Photovoltaic Cells, PhD Dissertation, University of Texas at Dallas, 2012.

[11] M. Levinshtein, S. Rumyantsev, M. Shur, Handbook Series on Semiconductor Parameters Volume 2: Ternary and Quaternary III–V Compounds, World Scientific, Singapore, 1999.

[12] B.A. Andersson, Materials availability for large-scale thin-film photovoltaics, Progress in Photovoltaics: Research and Applications 8 (2000) 61–76.

[13] A. Feltrin, A. Freundlich, Material considerations for terawatt level deployment of photovoltaics, Renewable Energy 33 (2008) 180–185.

[14] C.S. Tao, J. Jiang, M. Tao, Natural resource limitations to terawatt-scale solar cells, Solar Energy Materials and Solar Cells 95 (2011) 3176–3180.

[15] US DOE SunShot Initiative. Available from: http://energy.gov/eere/sunshot/sunshot-initiative.

[16] Sharp Electronics Corp. Available from: http://www.sharpsmartstorage.com.

CHAPTER *6*

Image Gallery

6.1 INTRODUCTION

Although not at all comprehensive in either scope or detail, some of the tangible products, processes, field installations, or tools related to the solar electricity marketplace and/or solar photovoltaic research and development are shown here (Figures 6.1–6.6). No specific product endorsement is implied.

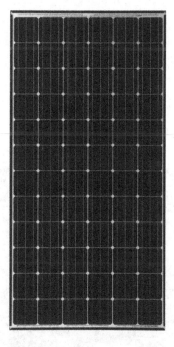

Fig. 6.1. Panasonic HIT® N245 module with 72 series-connected Si cells (heterojunction amorphous Si on monocrystalline Si). Technical specifications under standard test conditions (T_{cell} = 25°C and 1000 W m^{-2} AM1.5G spectrum): cell efficiency = 22.0%, module efficiency = 19.4%, maximum power = 245 W, maximum power voltage = 44.3 V, open circuit voltage = 53.0 V, maximum power current = 5.54 A, short circuit current = 5.86 A. Module dimensions = 1580 mm × 798 mm × 35 mm, module mass = 15 kg, glass material = tempered and AR coated, frame material = black anodized aluminum, power output guarantee = 10 years (90% of P_{min}) and 25 years (80% of P_{min}). Image courtesy of Panasonic Eco Solutions Energy Management Europe and Sanyo Component Europe GmbH.

Solar Photovoltaic Cells: Photons to Electricity. http://dx.doi.org/10.1016/B978-0-12-802329-7.00006-7

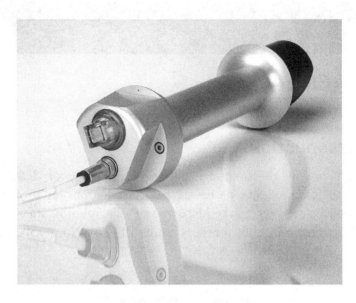

Fig. 6.2. Kipp & Zonen SHP1 pyrheliometer used to accurately measure DNI (direct normal irradiance). Technical specifications: spectral range = 200–4000 nm, maximum irradiance = 4000 W m^{-2}, operating temperature range = –40°C to 80°C, operating humidity range = 0–100% (relative humidity), expected daily uncertainty <1%, pyrheliometer mass = 0.9 kg (excluding cable). Note that global irradiance may be measured with a pyranometer that is not shown here. (Image courtesy of Kipp & Zonen B. V.)

Fig. 6.3. REC Silicon processing plant in Moses Lake, WA, USA. This plant has Siemens and fluidized bed reactor (FBR) technology for producing polycrystalline Si, a feedstock for multicrystalline and monocrystalline Si solar photovoltaic cells. From REC Silicon Inc.

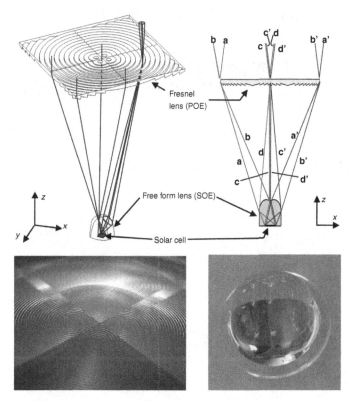

Fig. 6.4. Fresnel–Köhler primary–secondary optics that may be used for CPV solar applications to concentrate and focus sunlight onto small yet high efficiency multijunction solar photovoltaic cells [1]. (Image courtesy of Optics Express.)

Fig. 6.5. Semprius SM1 Series CPV system. Technical specifications: 14 kW$_{DC}$ (1000 W m^{-2} DNI, AM1.5D, T_{cell} = 25°C), 160 modules per tracker, 660 triple-junction (3J) cells per module, sunlight concentration ratio = 1111. Module dimensions = 636 mm × 476 mm × 68 mm, module mass = 7.3 kg, primary lens = silicone-on-glass (tempered), enclosure = powder-coated steel. (Image courtesy of Semprius Inc.)

Fig. 6.6. SMA Solar Technology Sunny Central 800CP–US utility grade inverter. Technical specifications: maximum DC power = 898 kW, maximum input voltage = 1000 V, nominal AC power at 50°C = 800 kVA, maximum efficiency = 98.7%, inverter dimensions = 2562 mm × 2272 mm × 956 mm, inverter mass < 1870 kg. (Image courtesy of SMA America, LLC.)

REFERENCE

[1] P. Benítez, J.C. Miñano, P. Zamora, R. Mohedano, A. Cvetkovic, M. Buljan, et al. High performance Fresnel-based photovoltaic concentrator, Optics Express 18 (2010) A25–A40.

Because solar photovoltaic cell technology is continually evolving in terms of metrics such as device open circuit voltage and power conversion efficiency, it is expected (and in fact desired) that some of the data contained in this book will fall out of date (e.g., Figure 3.7 that reported the published world record open circuit voltage for direct bandgap InP, GaAs, CdTe, and GaInP cells as of June 2014). As one example of ongoing innovation, Panasonic modified its HIT® technology by including an all-back contact electrode architecture to achieve a new world record Si solar photovoltaic cell with 25.6% power conversion efficiency (143.7 cm^2 designated illumination area) in February 2014 [1]. In August 2014, First Solar achieved 21.0% power conversion efficiency for a polycrystalline thin-film CdTe cell [2].

It is anticipated that new power conversion efficiency records exceeding 22% will soon be achieved with CdTe and CIGS cell technologies and 27% with monocrystalline Si cell technology (one Sun operation). In addition, it is anticipated that Si CPV cells will soon exceed 29% efficiency. More importantly, it is anticipated that CdTe and CIGS module efficiency (one Sun operation) will soon exceed 19%, and Si module efficiency (one Sun operation) will soon exceed 24%. Finally, it is anticipated that multijunction CPV modules will soon exceed 40% peak efficiency based on cells with four to six junctions.

As solar photovoltaic modules continue to scale up in deployment, the amount of solar electricity generation will increase. It is important to notice the difference between electricity generation and installed capacity; the latter is typically presented in discussions on solar electricity. In Chapter 1, the more meaningful metric of electricity generation was presented, rather than installed capacity.

It is anticipated that there will be fluctuation in the top 10 solar PV countries as to be expected when political and economic factors (such as feed-in-tariff incentives, etc.) are in play. This, in turn, should rapidly render the specific data in Figure 1.6 out of date. Nonetheless, this data

Solar Photovoltaic Cells: Photons to Electricity. http://dx.doi.org/10.1016/B978-0-12-802329-7.00001-8

could be thought of as an interesting historical benchmark that allows us to gauge how far we have progressed over the years.

The author will be just as curious as the reader as to how solar photovoltaic cell and module technology and the solar electricity marketplace evolve, beginning in 2014 on the 175th anniversary of the discovery of the photovoltaic effect.

REFERENCES

[1] M.A. Green, K. Emery, Y. Hishikawa, W. Warta, E.D. Dunlop, Solar cell efficiency tables (version 44), Progress in Photovoltaics: Research & Applications. 22 (2014) 701–710.

[2] B. Willis, First Solar hits 21.0% thin-film PV record. Available from: http://www.pv-tech.org/news/first_solar_hits_21.0_thin_film_pv_record.

FINAL REMARKS

It is important to note that solar photovoltaic cells and modules are not a miracle technology and that they can only be as "green" as we humans choose to design, develop, manufacture, and deploy them. To scale up solar PV widely beyond the near-trivial level of existence in 2014 will require a tremendous amount of raw material and large-scale manufacturing. Care must be taken to ensure that solar PV is done right (e.g., utmost efficiency in design and manufacturing at the front end and dedicated end-of-life recycling on the back end), or else the risk is that solar PV could end up posing its own set of legacy burdens on the planet.

A holistic dedication to energy efficiency in our daily life is perhaps the best way to enable solar photovoltaics to meet our demand for electricity. Inefficient incandescent and fluorescent lighting could be replaced with efficient LED and OLED lighting (equipped with motion sensors to turn off when not needed) as well as passive solar lighting in the daytime. More efficient air conditioning systems and heat pumps coupled with much better building insulation and better windows and improved door seals can reduce electricity demand. Likewise, improved efficiency of appliances such as refrigerators and industrial motors and the processes they run can reduce electricity demand. Using solar thermal energy to heat water for bathing and laundry eliminates the electricity required in the case of electric water heaters (and displaces the natural gas that would be needed for gas water heaters).

Lifestyle choices such as reduction (or outright elimination) of use of clothes dryers, hair blow dryers, and toaster ovens – all of which require so much electricity – could be another way to reduce electricity demand. Computers, e-readers, mobile telephones, and televisions (including set-top boxes) could be engineered to have radically reduced electricity demand compared to the inefficient products on the market in the early twenty-first century.

Overall, the less electricity that is needed in the first place, the less solar photovoltaic cells and modules that are needed, which then manifests as a reduced demand on semiconductors and balance-of-system

Solar Photovoltaic Cells: Photons to Electricity. http://dx.doi.org/10.1016/B978-0-12-802329-7.00008-0

(including energy storage system) component raw materials and the mining, processing, and environmental degradation that goes along with them. Therefore, energy efficiency is really the key for more sustainable and intelligent living, and the deployment of solar photovoltaic modules is synergistic with – and hopefully even guides and encourages – an energy efficient lifestyle.

APPENDIX A

List of Symbols

Å	Ångström (10^{-10} m)
A	vector potential
a	absorptivity
a	polarization vector
B	radiative recombination coefficient
c	velocity of light in vacuum
C	sunlight concentration factor
C_{max}	maximum sunlight concentration factor
C_n	Auger recombination coefficient of electrons
C_p	Auger recombination coefficient of holes
D	declination angle
$D(E)$	spectral photon density
D_n	diffusion coefficient of electrons
D_p	diffusion coefficient of holes
E	electric field
e	elementary charge (electron charge)
E	energy
E_C	bottom of conduction band
E_i	intrinsic Fermi level
E_F	Fermi energy level
E_{Fn}	quasi-Fermi level for electrons
E_{Fp}	quasi-Fermi level for holes
E_g	bandgap energy
E_V	top of valence band
F	free energy
FF	fill factor
G	gravitational constant
$g(E)$	electronic density of states

Solar Photovoltaic Cells: Photons to Electricity. http://dx.doi.org/10.1016/B978-0-12-802329-7.00009-2

$g_{ph}(E)$	photon density of states
$\mathbf{H}$	magnetic field
h	Planck constant
$\hbar$	reduced Planck constant ($h\,/\,2\pi$)
H'	perturbation Hamiltonian
I	current
I	irradiance
I_{AM0}	air mass zero total integrated irradiance
I_λ	spectral irradiance
I_m	maximum power point current
$I_{Rayleigh}$	Rayleigh scattering intensity
I_{sc}	short circuit current
J	current density
J_0	dark reverse saturation current density
J_m	maximum power point current density
J_{ph}	photogenerated current density
J_{rad}	radiative current density
J_{sc}	short circuit current density
$J–V$	current density–voltage
k_B	Boltzmann constant
L	latitude
L	luminosity (of the Sun)
L_n	diffusion length of electrons
L_p	diffusion length of holes
m^*	effective mass
m_e	electron mass
m_p	proton mass
m_S	mass of the Sun
n	electron concentration
$N(k,v)$	number of photon states
n_0	equilibrium electron concentration
N_A	acceptor doping concentration
N_C	effective density of conduction band states

N_D	donor doping concentration
n_i	intrinsic carrier concentration
n_{ph}	photogenerated electron concentration
n_r	index of refraction
N_T	trap concentration
N_V	effective density of valence band states
N_ω	number of photons
$\mathbf{p}$	momentum
p	hole concentration
P	atmospheric pressure
p_0	equilibrium hole concentration
P_0	standard atmospheric pressure
P_{in}	input power density (from the Sun)
P_{loss}	parasitic power loss
P_{out}	output power density (from the solar cell)
p_{ph}	photogenerated hole concentration
$\mathbf{q}$	photon wave vector
R	electrical resistance
$\mathbf{r}$	position
R_S	radius of the Sun
$\mathbf{S}$	Poynting vector
s	entropy per (photogenerated) electron (hole)
t	time
T	temperature
T_{eff}	effective surface temperature of Sun
T_{cell}	temperature of solar cell
T_{core}	core temperature of the Sun
$T_{hot\,carrier}$	hot carrier temperature
$T_{lattice}$	lattice temperature
TW_{peak}	peak power output in terawatts
V	volume
$\mathbf{v}$	velocity
V_{bi}	built-in voltage

V_m	maximum power point voltage
V_{oc}	open circuit voltage
v_{th}	thermal velocity
W_{abs}	absorption rate
z	thickness of solar cell absorber layer
α	Ångström turbidity exponent
$\alpha(E)$	absorption coefficient
β	Ångström turbidity coefficient
γ	gamma-ray photon
Δ	change in (free energy, chemical potential)
ε_0	vacuum permittivity
ε_{op}	optical permittivity
η	power conversion efficiency
θ	polar angle
θ	zenith angle
λ	photon wavelength
μ	chemical potential
μ	permeability
μ_n	electron drift mobility
μ_p	hole drift mobility
ν	photon frequency
ν_e	electron neutrino
ξ	azimuthal angle
π	pi
ρ	mass density
σ	Stefan–Boltzmann constant
σ_n	electron capture cross section
τ_{Aug}	Auger lifetime
τ_n	electron lifetime
τ_p	hole lifetime
τ_{rad}	radiative lifetime

τ_{SRH}	Shockley–Read–Hall lifetime
Φ	photon flux
$\varphi(\lambda, E)$	spectral photon flux
Ω	solid angle
Ω_{Sun}	solid angle subtended by the Sun
ω	photon angular frequency ($2\pi\nu$)

APPENDIX *B*

Abbreviations and Acronyms

III–V	compound semiconductor comprised of group III and V elements
1J	single junction (solar cell)
2J	double junction (solar cell)
3J	triple junction (solar cell)
4J	four-junction (solar cell)
6J	six-junction (solar cell)
9J	nine-junction (solar cell)
a–Si:H	amorphous silicon (hydrogenated)
AC	alternating current
A/C	air conditioning
Ag	silver
Al_2O_3	aluminum oxide
AlGaAs	aluminum gallium arsenide
AlGaInAs	aluminum gallium indium arsenide
AlGaInP	aluminum gallium indium phosphide
AlInP	aluminum indium phosphide
As	arsenic
AM	air mass
AM0	air mass zero
AM1	air mass 1
AM1.5	air mass 1.5
AM1.5D	air mass 1.5 direct
AM1.5G	air mass 1.5 global
AM2	air mass 2
AM38	air mass 38
$AM_{(min)}$	minimum air mass
amu	atomic mass unit

Solar Photovoltaic Cells: Photons to Electricity. http://dx.doi.org/10.1016/B978-0-12-802329-7.00010-9

AOD	aerosol optical depth
ARC	antireflection coating
As	arsenic
ASTM	American Society for Testing and Materials
AU	astronomical unit
AZ	Arizona
BP	British Petroleum
BSF	back surface field
BZ	Brillouin zone
c–Si	crystalline silicon
CB	conduction band
Cd	cadmium
CdS	cadmium sulfide
CdTe	cadmium telluride
CFL	compact fluorescent lamp
CGS	copper gallium diselenide
CH_4	methane
CH_2O	carbohydrate subunit
CIGS	copper indium gallium diselenide
CIS	copper indium diselenide
CMP	chemical mechanical polishing
CO	carbon monoxide
CO_2	carbon dioxide
CSP	concentrated solar power
CPV	concentrated photovoltaic
Cu	copper
CZTSS	copper zinc tin sulfide selenide
DB	detailed balance
DBR	distributed Bragg reflector
DC	direct current
DNI	direct normal irradiance
DOE	Department of Energy (U.S.)
EM	electromagnetic

EOL	end-of-life
ESC	energy selective contact
ESR	electrical substitution radiometer
EU	European Union
FBR	fluidized bed reactor
FF	fill factor
Ga	gallium
GaAs	gallium arsenide
GaInAs	gallium indium arsenide
GaInN	gallium indium nitride
GaInNAsSb	gallium indium nitride arsenide antimonide
GaInP	gallium indium phosphide
GaN	gallium nitride
Ge	germanium
GeSbTe	germanium–antimony–tellurium
GNI	global normal irradiance
Hα	hydrogen–alpha
H_2O	water
HCl	hydrogen chloride
hep	helium–proton (neutrino)
Hg	mercury
HH	heavy hole (valence band)
HIT	heterojunction with intrinsic thin layer
HRSG	heat recovery steam generator
IGZO	indium gallium zinc oxide
In	indium
InP	indium phosphide
IR	infrared
ITER	International Thermonuclear Experimental Reactor
J–V	current density–voltage
K	Kelvin (temperature)
kcal	kilocalorie

kg	kilogram
kWh	kilowatt–hour
LED	light-emitting diode
LH	light hole (valence band)
Li	lithium
LNG	liquefied natural gas
LO	longitudinal optical (phonon)
MBE	molecular beam epitaxy
MgF_2	magnesium fluoride
MIS	metal–insulator–semiconductor
MN	Minnesota
MOVPE	metal organic vapor phase epitaxy
NASA	National Aeronautics and Space Administration
NIF	National Ignition Facility
Ni–P	nickel–phosphorus
NO_2	nitrogen dioxide
NRL	Naval Research Laboratory
O_2	oxygen
O_3	ozone
OLED	organic light emitting diode
pep	proton–electron–proton (neutrino)
PMT	photomultiplier tube
pp	proton–proton (neutrino or chain)
PV	photovoltaic
PWh	petawatt–hour
Sb	antimony
Se	selenium
Si	silicon
$SiCl_4$	silicon tetrachloride
SiGe	silicon germanium
SiH_4	silane
SiH_2Cl_2	dichlorosilane
$SiHCl_3$	trichlorosilane

SiN_x:H	silicon nitride (hydrogenated)
SiO_2	silicon dioxide
SK	Super–Kamiokande
SMARTS	Simple Model of the Atmospheric Radiative Transfer of Sunshine
SNO	Sudbury Neutrino Observatory
SO	split-off valence band
SO_2	sulfur dioxide
SORCE	Solar Radiation and Climate Experiment (satellite)
SRH	Shockley–Read–Hall (recombination)
TCTE	total solar irradiance calibration transfer experiment (satellite)
Te	tellurium
TEPCO	Tokyo Electric Power Company
TIM	total irradiance monitor
TiN	titanium nitride
TO	transverse optical (phonon)
TSI	total solar irradiance
TSI_{avg}	average total solar irradiance
TW	terawatt
^{235}U	uranium 235 (fissile isotope)
UN	United Nations
USA	United States of America
USGS	United States Geological Survey
UV	ultraviolet
VB	valence band
W	Watt
W	tungsten
Wh	watt–hour
Zn	zinc
ZnO	zinc oxide
ZnS	zinc sulfide

APPENDIX C

Physical Constants

Astronomical unit	AU	1.496×10^{11} m
Boltzmann constant	k_B	1.3806×10^{-23} J K^{-1}
Electron mass	m_e	9.109×10^{-31} kg
Elementary charge	e	1.60218×10^{-19} C
Gravitational constant	G	6.6738×10^{-11} N m^2 kg^{-2}
Mass of Sun	m_S	1.99×10^{30} kg
Planck constant	h	6.6262×10^{-34} J s
Planck constant (reduced)	$\hbar$	1.0546×10^{-34} J s
Proton mass	m_p	1.6726×10^{-27} kg
Radius of Sun	R_S	6.96×10^8 m
Solid angle subtended by Sun	Ω_{Sun}	6.8×10^{-5} sr
Standard atmospheric pressure	P_0	1.01325×10^5 N m^{-2}
Stefan–Boltzmann constant	σ	5.67×10^{-8} W m^{-2} K^{-4}
Vacuum permittivity	ε_0	8.854×10^{-12} A^2 s^4 kg^{-1} m^{-3}
Velocity of photons in vacuum	c	2.9979×10^8 m s^{-1}

Solar Photovoltaic Cells: Photons to Electricity. http://dx.doi.org/10.1016/B978-0-12-802329-7.00011-0

APPENDIX *D*

Conversion Factors

1 C (Coulomb)	$= 1 \, A \, s$
1 eV (electron volt)	$= 1.60218 \times 10^{-19} \, J$
1 J (Joule)	$= 1 \, W \, s = 1 \, kg \, m^2 \, s^{-2}$
1 kWh (kilowatt–hour)	$= 859.85 \, kcal$
K (Kelvin)	$= °C + 273.15$
λ (nm)	$= 1239.85 \, / \, E \, (eV)$

Solar Photovoltaic Cells: Photons to Electricity. http://dx.doi.org/10.1016/B978-0-12-802329-7.00012-2

Derivation of Absorption Coefficient

step 1

When a photon (EM radiation) interacts with an electron in, for example, a semiconductor lattice, a perturbation occurs meaning that the initially unperturbed electron (e.g., in a valence band quantum state) may be excited to an empty quantum state (e.g., in a conduction band, if the photon has energy equal to or greater than the semiconductor bandgap). Here, we will follow, in part, the analysis from B. K. Ridley, *Quantum Processes in Semiconductors, Fifth Edition*, Oxford University Press, 2013. The perturbation is treated quantum mechanically with what is known as a perturbation Hamiltonian H', given by:

$$H' = -\frac{e}{m_e} \mathbf{A} \cdot \mathbf{p}, \qquad (E.1)$$

where e is the elementary (electron) charge, m_e is the mass of the electron, $\mathbf{A}$ is the vector potential (we assume $\nabla \cdot \mathbf{A} = 0$), and $\mathbf{p}$ is the momentum operator. The vector potential is expressed as:

$$\mathbf{A} = A_0 \, \mathbf{a} \cos(\mathbf{q} \cdot \mathbf{r} - \omega t), \qquad (E.2)$$

where $\mathbf{a}$ is the polarization vector, $\mathbf{q}$ is the photon wave vector, $\mathbf{r}$ is position, ω is the photon angular frequency, and t is time.

step 2

Next, we may determine the Poynting vector $\mathbf{S}$ expressed as:

$$\mathbf{S} = \mathbf{E} \times \mathbf{H}, \qquad (E.3)$$

where the electric field $\mathbf{E}$ is given by:

$$\mathbf{E} = -\frac{\partial \mathbf{A}}{\partial t} = -\omega A_0 \, \mathbf{a} \sin(\mathbf{q} \cdot \mathbf{r} - \omega t), \qquad (E.4)$$

Solar Photovoltaic Cells: Photons to Electricity. http://dx.doi.org/10.1016/B978-0-12-802329-7.00013-4

and the magnetic field **H** is given by:

$$\mathbf{H} = \frac{1}{\mu}\nabla \times \mathbf{A} = -\frac{1}{\mu}A_0(\mathbf{q}\times\mathbf{a})\sin(\mathbf{q}\cdot\mathbf{r}-\omega t), \tag{E.5}$$

where μ is the permeability. Substituting Equation E.4 and Equation E.5 into Equation E.3 gives:

$$\mathbf{S} = \frac{\omega A_0^2}{\mu}\mathbf{q}\sin^2(\mathbf{q}\cdot\mathbf{r}-\omega t). \tag{E.6}$$

The time average Poynting vector $\langle\mathbf{S}\rangle$ is given by:

$$\langle\mathbf{S}\rangle = \frac{A_0^2 q^2}{2\mu}\mathbf{v}, \tag{E.7}$$

where **v** is the velocity $(\omega = \mathbf{q}\cdot\mathbf{v})$. Now, the radiation energy density is given by:

$$\langle E\rangle = \frac{\langle\mathbf{S}\rangle}{\mathbf{v}} = \frac{A_0^2 q^2}{2\mu}, \tag{E.8}$$

where $\langle E\rangle = N_\omega\hbar\omega/V$ and N_ω is the number of photons while V is the semiconductor crystal volume. The magnitude of the velocity is given by $(\mu\varepsilon_{op})^{-1/2}$, where ε_{op} is the optical permittivity. Solving for A_0^2 in Equation E.8, we arrive at:

$$A_0^2 = \frac{2\hbar N_\omega}{V\varepsilon_{op}\omega}. \tag{E.9}$$

step 3

Next, Equation E.2 is rewritten as:

$$\mathbf{A} = \frac{A_0\mathbf{a}}{2}\{\exp[i(\mathbf{q}\cdot\mathbf{r}-\omega t)]+\exp[-i(\mathbf{q}\cdot\mathbf{r}-\omega t)]\}. \tag{E.10}$$

The first exponential term in Equation E.10 is related to photon absorption while the second exponential term is related to photon emission; we are only interested in the first exponential term here. Therefore, after dropping the second exponential term, we now arrive at the expression for the perturbation Hamiltonian H' by substituting the square root of Equation E.9 into Equation E.10 and then Equation E.10 into Equation E.1 resulting in:

$$H' = -\frac{e}{m_e}\left(\frac{\hbar N_\omega}{2V\varepsilon_{op}\omega}\right)^{1/2} \exp(i\mathbf{q}\cdot\mathbf{r})\mathbf{a}\cdot\mathbf{p}. \tag{E.11}$$

step 4

Considering direct interband transitions from the valence band (initial quantum state) to conduction band (final quantum state) where we have approximate conservation of electron wave vector $\mathbf{k}_i \approx \mathbf{k}_f$, we may now invoke the Fermi golden rule (which is an approximation) to determine, for a first-order time-dependent perturbation, the photon absorption rate $W_{abs}(E)$ for photons with energy $E \geq E_g$ as:

$$W_{abs}(E) = \frac{2\pi}{\hbar}\int dE_i g_i(E_i)\int dE_f g_f(E_f)V\,|\langle f|H'|i\rangle|^2 \\ \delta(E_f - E_i - E)[f(E_i) - f(E_f)], \tag{E.12}$$

where $|i\rangle$ and $\langle f|$ are the initial and final states, respectively, V is the semiconductor crystal volume, $g_i(E_i)$ and $g_f(E_f)$ are the initial and final electronic density of states respectively, and $f(E_i)$ and $f(E_f)$ are the Fermi occupation factors. After substituting Equation E.11 into Equation E.12 while invoking the dipole approximation, $\exp(i\mathbf{q}\cdot\mathbf{r}) \approx 1$, we obtain:

$$W_{abs}(E) = \frac{\pi e^2 N_\omega}{\varepsilon_0 n_r^2 m_e^2 \omega}\int|\langle f|\mathbf{a}\cdot\mathbf{p}|i\rangle|^2 g_i(E_i)g_f(E - E_g - E_i) \\ [f(E_i) - f(E - E_g - E_i)]dE_i, \tag{E.13}$$

where E is the photon energy and E_g is the bandgap energy. Note that we used $\varepsilon_{op} = \varepsilon_0 n_r^2$, where ε_0 is the vacuum permittivity and n_r is the index of refraction of the semiconductor.

step 5

Finally, if we assume a full valence band and empty conduction band, then the absorption coefficient $\alpha(E)$ may be found by dividing the absorption rate in Equation E.13 by the velocity of the photons in the semiconductor (c / n_r) and by the number of photons N_ω, resulting in:

$$\alpha(E) = \frac{\pi e^2}{c \varepsilon_0 n_r m_e^2 \omega} \int |\langle f | \mathbf{a} \cdot \mathbf{p} | i \rangle|^2 \, g_i(E_i) g_f(E - E_g - E_i) \, dE_i, \qquad (E.14)$$

where $\int g_i(E_i) g_f(E - E_g - E_i) \, dE_i$ is the joint electronic density of states. Note that the absorption coefficient for indirect bandgap semiconductors (such as Si) is derived in the framework of more complex second order time-dependent perturbation theory that has not been discussed here.

Derivation of Open Circuit Voltage

step 1

Electrons and holes radiatively recombine in a semiconductor layer of an illuminated solar photovoltaic cell via interband transitions (i.e., conduction to valence band transitions separated by the discrete energy bandgap E_g). These emitted photons are known as luminescent photons. The emitted luminescent photon flux Φ outgoing from the solar photovoltaic cell is given by:

$$\Phi = \exp(\Delta\mu / k_B T) \int_{E_g}^{\infty} a(E)\varphi(E)\,dE \int_{0}^{\pi/2} \cos\theta \sin\theta\,d\theta \int_{0}^{2\pi} d\xi, \quad \text{(F.1)}$$

where $\Delta\mu$ is change in chemical potential (i.e., difference in the electron and hole gas chemical potentials or quasi-Fermi levels in a perturbed semiconductor such as a solar photovoltaic cell under sunlight illumination; we will assume that the separation of quasi-Fermi levels is constant), k_B is Boltzmann's constant, T is cell temperature, E is photon energy, $a(E)$ is absorptivity, $\varphi(E)$ is spectral photon flux, θ is polar angle, and ξ is azimuthal angle.

step 2

Now, to eventually find $\varphi(E)$ – the emitted spectral photon flux – shown in the integrand of Equation F.1, we first need to determine the photon density of states. Here, we will follow, in part, the analysis from H. C. Casey, Jr. and M. B. Panish, *Heterostructure Lasers Part A: Fundamental Principles*, Academic Press, 1978. We consider a real-space cube of volume $V = L^3$. Invoking periodic boundary conditions, we find $k_x = k_y = k_z = 2\pi / L$. Therefore, the k-space volume is $(2\pi / L)^3$. The incremental number of photon states is given by:

$$dN(k) = 2\frac{4\pi k^2 dk}{(2\pi / L)^3}, \quad \text{(F.2)}$$

where the two different photon polarization states are represented by the value of 2, the volume of a spherical shell is represented by $4\pi k^2 dk$, the wave vector is represented by $k = 2\pi n_r v / c$ where n_r is index of refraction, v is photon frequency, c is photon velocity in vacuum, and finally $dk = (2\pi n_r / c)dv$. We make the substitutions and rewrite Equation F.2 as:

$$dN(v) = (L^3)8\pi n_r^3 v^2 c^{-3} dv. \tag{F.3}$$

Upon substituting $v = E / h$ and $dv = dE / h$, where h is Planck's constant, into Equation F.3, the photon density of states (i.e., number of states per unit volume per unit energy) is given by:

$$g_{ph}(E) = dN(E)/VdE = 8\pi n_r^3 E^2 h^{-3} c^{-3}. \tag{F.4}$$

Now, we can determine the photon density of states per unit solid angle. Photons are emitted in an isotropic manner, or over a full solid angle of $\Omega = 4\pi$. Therefore, the photon density of states per solid angle is found by dividing Equation F.4 by 4π, which results in:

$$g_{ph,\Omega}(E) = 2n_r^3 E^2 h^{-3} c^{-3}. \tag{F.5}$$

step 3

The Bose–Einstein distribution function expresses the probability that photons (bosons) will be in a given energy state as:

$$f(E) = [\exp(E / k_B T) - 1]^{-1} \tag{F.6}$$

By invoking the "classical" regime, we may assume that $\exp(E/k_B T) \gg 1$, and therefore, Equation F.6 is rewritten as:

$$f(E) \approx \exp(-E / k_B T). \tag{F.7}$$

The spectral photon density is found by multiplying Equation F.5 by Equation F.7, resulting in:

$$D(E) = 2n_r^3 E^2 h^{-3} c^{-3} \exp(-E / k_B T). \tag{F.8}$$

The spectral photon density $D(E)$ is converted to spectral photon flux $\varphi(E)$ by multiplying Equation F.8 by the photon velocity c / n_r.

Therefore, the spectral photon flux (that is shown in the integrand of Equation F.1) is given by:

$$\varphi(E) = 2n_r^2 E^2 h^{-3} c^{-2} \exp(-E / k_B T). \tag{F.9}$$

step 4

Next, we will determine the radiative current density J_{rad}. To do this, we need to revert to Equation F.1 and multiply the emitted luminescent photon flux Φ by the electronic charge, e, which gives:

$$J_{rad} = e\left[\exp(\Delta\mu / k_B T) \int_{E_g}^{\infty} a(E)\varphi(E)dE \int_0^{\pi/2} \cos\theta \sin\theta \, d\theta \int_0^{2\pi} d\xi\right]. \tag{F.10}$$

We shall consider the ideal case for a solar photovoltaic cell in which $a(E) = 1$, which means unity absorptivity for photons with energy equal to or greater than the bandgap energy ($E \geq E_g$). We consider the presence of a perfect backside mirror so that the emitted luminescent photons only escape out of the front surface of the cell, noting that $n_r = 1$ for air. Now, solving the integrals in Equation F.10, we find:

$$\int_0^{\pi/2} \cos\theta \sin\theta \, d\theta \int_0^{2\pi} d\xi = \pi, \tag{F.11}$$

and

$$\int_{E_g}^{\infty} 2E^2 h^{-3} c^{-2} \exp(-E / k_B T) dE$$
$$= 2k_B T h^{-3} c^{-2} \exp(-E_g / k_B T)[E_g^2 + 2(k_B T)E_g + 2(k_B T)^2]. \tag{F.12}$$

Since the terms $2(k_B T)E_g$ and $2(k_B T)^2$ are much smaller than E_g^2 in Equation F.12, they are neglected in the interest of keeping only the dominant terms. Therefore, Equation F.10 for the radiative current density reduces to:

$$J_{rad} = \exp(\Delta\mu / k_B T) 2\pi e E_g^2 k_B T h^{-3} c^{-2} \exp(-E_g / k_B T). \tag{F.13}$$

step 5

In an ideal solar photovoltaic cell at steady state and open circuit and considering the absence of any nonradiative recombination, the photogenerated current density J_{ph} (incoming absorbed solar photon flux multiplied by the electronic charge) balances the radiative current density J_{rad} (outgoing emitted luminescent photon flux multiplied by the electronic charge). This condition, explained by the principle of detailed balance for constant quasi-Fermi level separation, is represented by:

$$J_{ph} = J_{rad}. \tag{F.14}$$

Therefore, substituting Equation F.13 into Equation F.14, we find:

$$J_{ph} = \exp(\Delta\mu / k_B T) 2\pi e E_g^2 k_B T h^{-3} c^{-2} \exp(-E_g / k_B T). \tag{F.15}$$

The change in chemical potential for an illuminated solar photovoltaic cell at the open circuit condition is expressed as $\Delta\mu = e V_{oc}$, and therefore, we rewrite Equation F.15 as:

$$J_{ph} = \exp(e V_{oc} / k_B T) 2\pi e E_g^2 k_B T h^{-3} c^{-2} \exp(-E_g / k_B T). \tag{F.16}$$

Finally, following mathematical manipulation of Equation F.16, we arrive at the following expression for the detailed balance-limiting open circuit voltage V_{oc} in an ideal solar photovoltaic cell with a perfect backside mirror,

$$V_{oc} = E_g e^{-1} - k_B T e^{-1} \ln(2\pi e E_g^2 k_B T h^{-3} c^{-2} J_{ph}^{-1}). \tag{F.17}$$

Written in this specific format, Equation F.17 offers the most intuitive understanding of the fundamental factors that govern V_{oc} – the bandgap energy E_g, cell temperature T, and photogenerated current density J_{ph}. Note that for a bifacial solar photovoltaic cell, the emitted luminescent radiation may escape from the front (top) surface of the cell into air or escape out of the rear (bottom) surface of the cell into air. Therefore, a factor of 2 would be included in the logarithm term in Equation F.17. If the solar photovoltaic cell instead contains a parasitically absorbing substrate (with no photovoltaic activity) on the rear, then the emitted luminescent radiation may escape from the front (top) surface of the cell into air or escape out of the rear (bottom) surface of the active region of the cell where it is absorbed in the parasitic substrate below. Therefore, in this case, a factor of $(1 + n_r^2)$ would be included in the logarithm term in Equation F.17, where n_r is the refractive index of the parasitically absorbing substrate. Ultimately, an ideal mirror on the back of the solar photovoltaic cell offers the optimal limiting V_{oc}.

APPENDIX G

Relative Efficiency Ratio

To make a quick comparison of solar photovoltaic cell technologies based on amorphous (a–), polycrystalline (poly), multicrystalline (multi), and monocrystalline (mono) semiconductors, it is helpful to look at the world record cell power conversion efficiency [1,2] divided by the detailed balance-limiting power conversion efficiency as a function of the particular solar photovoltaic cell's bandgap energy. In this case, the detailed balance limit for cells containing an ideal backside mirror (reflector) is invoked because the mirror configuration gives the greatest power conversion efficiency as discussed in Chapter 3. Relative efficiency ratio of various single junction cell technologies (as of August 2014) is shown in Figure G.1.

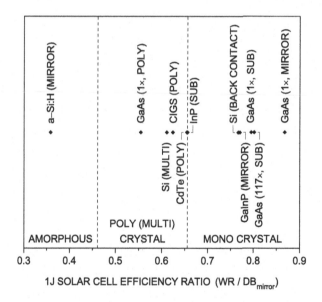

Fig. G.1. Relative efficiency ratio of a–Si:H (E_g = 1.75 eV), poly-GaAs, multi-Si, poly-CIGS (E_g = 1.15 eV), poly-CdTe, mono-InP, mono-Si, mono-GaInP, and mono-GaAs. Indirect bandgap (multi and mono) Si is assumed to be radiatively limited. Note that poly-CdTe has matched the relative efficiency ratio of mono-InP, while the record one Sun GaAs (mirror) cell has outperformed the record 117× GaAs cell (a substrate variant). As of August 2014, no 1J solar cell had exceeded a relative efficiency ratio of 0.9. Key: SUB is substrate, WR is world record power conversion efficiency, and DB$_{mirror}$ is the detailed balance-limiting power conversion efficiency assuming the cells have a perfect backside mirror under the AM1.5 spectrum and 298 K cell temperature. (After Ref. [3].)

Solar Photovoltaic Cells: Photons to Electricity. http://dx.doi.org/10.1016/B978-0-12-802329-7.00015-8

REFERENCES

[1] M.A. Green, K. Emery, Y. Hishikawa, W. Warta, E.D. Dunlop, Solar cell efficiency tables (version 44), Progress in Photovoltaics: Research and Applications 22 (2014) 701–710 .

[2] B. Willis, First Solar hits 21.0% thin-film PV record. Available from: http://www.pv-tech.org/news/first_solar_hits_21.0_thin_film_pv_record.

[3] A.P. Kirk, M.J. DiNezza, S. Liu, X.–H. Zhao, Y.–H. Zhang. CdTe vs. GaAs solar cells – a modeling case study with preliminary experimental results, in: Proc. IEEE 39th Photovoltaic Specialists Conference, 2013, pp. 2515–2517.

Recalibrating the Orthodoxy

Some examples of the existing orthodoxy concerning solar photovoltaic cells and their operation are presented below. Corrections, clarifications, and/or comments are then presented in an attempt to recalibrate how we think about solar photovoltaic cells.

Assertion 1

The built-in voltage V_{bi} of a p–n junction solar photovoltaic cell in the dark determines the open circuit voltage V_{oc} of the cell under sunlight illumination.

The device design engineer fixes the dark-state built-in voltage V_{bi} of a p–n junction (diode) via choice of acceptor and donor doping concentration. To be clear, there is no rule that governs the exact doping concentration that should be chosen when designing a solar photovoltaic cell. The expression for V_{bi} (nondegenerate) is given by:

$$V_{bi} = k_B T e^{-1} \ln(N_A N_D / n_i^2). \tag{H.1}$$

Let us consider a GaAs p–n junction solar photovoltaic cell as an example. From prior detailed balance arguments discussed in Chapter 3 and Appendix F, a 1.42 eV GaAs cell with an ideal backside mirror has a maximum $V_{oc} = 1.157$ V (AM1.5G 1× spectrum, $T_{cell} = 300$ K). Note that detailed balance calculations are made independent of doping concentration. Now, if we assume nondegenerate acceptor doping concentration ($N_A = 4 \times 10^{17}$ cm^{-3}) and nondegenerate donor doping concentration ($N_D = 2 \times 10^{16}$ cm^{-3}) in a GaAs p–n junction cell ($n_i = 2.36 \times 10^6$ cm^{-3} at 300 K), then from Equation H.1, we find that $V_{bi} = 1.259$ V at 300 K, which well exceeds the thermodynamically governed detailed balance-limiting V_{oc} at 300 K. Therefore, the V_{bi} describing the p–n junction in the dark should not be confused with the V_{oc} of a solar photovoltaic cell under sunlight illumination.

Solar Photovoltaic Cells: Photons to Electricity. http://dx.doi.org/10.1016/B978-0-12-802329-7.00016-X

Assertion 2

The ideal diode equation provides an intuitive way to understand the relationship between the thickness z and open circuit voltage V_{oc} of a solar photovoltaic cell.

The ideal diode equation open circuit voltage V_{oc} is given by:

$$V_{oc} = k_B T e^{-1} \ln(1 + J_{ph} / J_0), \qquad (H.2)$$

where J_{ph} is the photogenerated current density and J_0 is the dark reverse saturation current density. From Equation H.2, it appears that a reduction in J_0 will give an increase in V_{oc}. The usual argument is that an arbitrary reduction in a solar cell's thickness, z, will manifest as a reduction in J_0 resulting in an increase in V_{oc} while an arbitrary increase in a solar cell's thickness, z, will manifest instead as an increase in J_0 resulting in a reduction in V_{oc}. Let us again consider GaAs p–n junction solar photovoltaic cells as an example. In one case, we have a 4 μm thick GaAs cell with no mirror, and in the other case, we have a 2 μm thick GaAs cell with a perfect backside mirror, which means the effective optical depth of this cell is the same as the 4 μm mirror-free cell. Both cells have the same area (e.g., 1 cm²); therefore, the physical volume of the 2 μm thick cell with the mirror is half that of its 4 μm thick counterpart. Let us also assume both cells are able to absorb the same number of photons (due to the same effective optical depth, same absorption coefficient, and same area), and therefore, each cell has the same photogenerated current density, J_{ph}. Considering actual solar photovoltaic cells, the reason that a thinner GaAs cell can have a larger V_{oc} (but not larger than the detailed balance-limiting V_{oc} that was calculated independent of thickness) is due to the reduced entropy production per photogenerated carrier. Entropy production leads to V_{oc} degradation. From the Sackur–Tetrode expression for an ideal (nondegenerate) Fermi gas, the entropy per photogenerated electron, s, is given by:

$$s \propto k_B[5/2 - \ln(n_{ph} / N_C)], \qquad (H.3)$$

where n_{ph} is the photogenerated electron concentration and N_C is the effective conduction band density of states. Since voltage is a function of entropy, from Equation H.3, we see that an increase in n_{ph} results in a reduction in s, and therefore an enhanced V_{oc} because the thinner cell (that was able to achieve the same J_{ph} as the thicker cell) has a smaller volume and therefore larger n_{ph}. The increase in voltage as a function of an increase in photogenerated carrier concentration can be seen from the following relationship given by:

$$V_{oc} \propto k_B T e^{-1} \ln(n_{ph} p_{ph} / n_i^2). \tag{H.4}$$

The analysis presented here was previously explained with more detail by Brendel and Queisser whereby they showed that consideration of the entropy production per photogenerated carrier provides a more intuitive understanding of the relationship between cell thickness (volume) and open circuit voltage V_{oc} compared to the ideal diode equation and its dark reverse saturation current density J_0 [1]. Repeating the earlier caveat, the detailed balance limit sets the maximum V_{oc} ceiling because the detailed balance calculations are made independent of cell thickness and doping concentration.

Assertion 3

Luminescent photons are recycled inside a solar photovoltaic cell.

From quantum mechanics, there are creation and annihilation operators for bosons and fermions [2]. When an electron and hole recombine radiatively in an interband transition in a semiconductor, a luminescent photon is created (emitted) and the electron and hole, fermions, are annihilated. This luminescent photon, a boson, may be absorbed in the semiconductor. If so, the luminescent photon is annihilated resulting in a photogenerated electron in the conduction band and a hole in the valence band. Luminescent photon emission and absorption is a natural process in semiconductors, including for unprocessed bare wafers of Si or GaAs sitting on a lab bench and illuminated by a desk lamp. The main point here is that while there certainly is luminescent photon emission and absorption, there is no actual recycling (or reuse) of any given photon because absorption of the photon results in its annihilation. In addition, in most inorganic semiconductors at room temperature, a photogenerated carrier pair (known as an exciton) rapidly becomes a free electron and free hole; these mobile free carriers diffuse and/or drift prior to radiative (or nonradiative) recombination. Therefore, a photogenerated electron does not necessarily recombine with the exact same hole that was created upon absorption of the luminescent photon.

Assertion 4

The principle of detailed balance only applies to the condition of dark-state thermal equilibrium where the solar photovoltaic cell absorbs and emits thermal radiation and the cell is described as having a single Fermi level and therefore voltage equal to zero.

The principle of detailed balance (where we consider radiative recombination as the only carrier recombination mechanism) also applies to sunlight-illuminated solar photovoltaic cells at steady-state open circuit condition with constant quasi-Fermi level separation [3]. In the radiative limit, the absorbed photogenerated current density must balance the emitted radiative current density (see Appendix F). Each absorbed solar photon generates one electron and one hole. The photogenerated (i.e., excited) electrons cannot remain in the conduction band indefinitely; therefore, in the radiatively limited balance at steady-state and open circuit, one radiative recombination event involves one electron and one hole radiatively recombining to yield one luminescent photon.

Assertion 5

Multijunction solar photovoltaic cells exceed the detailed balance limit.

Although multijunction solar photovoltaic cells are more efficient than standalone single junction cells, they are comprised of individual single junction cells known as subcells (typically configured in a monolithic series connection by using tunnel diodes) and therefore are governed by the same detailed balance formalism in the radiative limit that Shockley and Queisser employed for their specific analysis of standalone single junction cells [4,5]. Refer to Chapters 3 and 4 for more information about multijunction cells and examples of their limiting efficiency according to the principle of detailed balance.

Assertion 6

Hot-carrier solar photovoltaic cells and intermediate band solar photovoltaic cells (i.e., the so-called "3rd Generation" photovoltaic devices) offer superior power conversion efficiency when compared to multijunction solar photovoltaic cells.

As the number of subcells approaches infinity, a multijunction solar photovoltaic cell of this type offers the maximum possible power conversion efficiency for a photovoltaic device because (1) an infinite number of subcells can be tuned to absorb all of the solar photons in, for example, the AM1.5 G (AM1.5D) spectrum from 280 nm (~4.43 eV) to 4000 nm (~0.31 eV), and (2) since the individual subcells would be configured to respond in the monochromatic limit, this means that any given subcell only absorbs photons with energy equivalent to the subcell's bandgap energy and therefore this in turn means that there is no hot carrier relaxation loss. As such, if we are strictly considering a solar photovoltaic device, then the infinite subcell multijunction design yields the maximum possible photovoltaic power conversion efficiency. Practically speaking, a well-designed 6-junction cell (such as the lattice-matched configuration discussed in Chapter 4) already achieves close enough to the maximum limiting efficiency that there is no need to be obsessed with a greater number of subcells for a multijunction architecture. Additionally, a sunlight concentration factor of ~1000× is adequate since the incremental gain in power conversion efficiency beyond 1000× is small, and therefore, there is no need to be obsessed with increasing sunlight concentration toward the limit of ~46200×.

REFERENCES

[1]   R. Brendel, H.J. Queisser, On the thickness dependence of open circuit voltages of p-n junction solar cells, Solar Energy Materials and Solar Cells 29 (1993) 397–401.

[2]   H. Kroemer, Quantum Mechanics for Engineering, Materials Science, and Applied Physics, Prentice Hall, Englewood Cliffs, 1994.

[3]   G.L. Araújo, A. Martí, Absolute limiting efficiencies for photovoltaic energy conversion, Solar Energy Materials and Solar Cells 33 (1994) 213–240.

[4]   W. Shockley, H.J. Queisser, Detailed balance limit of efficiency of p-n junction solar cells, Journal of Applied Physics 32 (1961) 510–519.

[5]   C.H. Henry, Limiting efficiencies of ideal single and multiple energy gap terrestrial solar cells, Journal of Applied Physics 51 (1980) 4494–4500.

Printed in the United States
By Bookmasters